ETHICS BASED ON THE SCIENCE OF EVOLUTION: NATURE + NURTURE

By

ROBERT GORDON, PhD

ISBN: 1-4033-0306-1 (e-book)
ISBN: 1-4033-0307-X (Paperback)

This book is printed on acid free paper.

1stBooks - rev. 04/04/03

TABLE OF CONTENTS

Page No.

Appendices

Figures

iv

INTRODUCTION

In formulating and presenting an ethical system based on the science of evolution, instead of religion or philosophy, it would appear to be appropriate to use the methods of science.

Thus, the terms to be used are defined, references are given and the methodology of mathematics is employed. However, this will be done in a simple way and the reader will not have to know any formal science or mathematics beyond high school.

After defining the initial terms to be used and some elements of the scientific method, a synopsis is given which explains the whole ethical system based on the science of evolution in simple terms without elaboration, references or justifications, all of which will be presented in the subsequent parts and chapters. With the structure, or "bare bones" of the proposed ethical system in mind, then the subsequent materials are presented in that concept, fleshing out the "bare bones." Statements that are not supported by factual evidence are labeled as opinions or conjecture.

The words ethics and morals, as defined in various dictionaries overlap and are defined in loose terms. These two words, ethics and morals, will be used in this book as follows:

a. Both ethics and morals are descriptions of human behavior. Such behavior may deemed to be "good" or "bad," or "right" or "wrong." These last four terms require definitions, which will be provided later in the context of the proposed system of scientific ethics.
b. Morals are based on religious precepts from a supernatural being.
c. Ethics are based on the reasoning of a philosopher.
d. Ethics based on science, as described later, will be called SciEthics in order to distinguish them from the ethics defined above.

It is acknowledged that the above definitions are not exactly as defined by Webster, but they are given here explicitly, so that the reader will not have to guess at the meanings used by the author. Defining terms used is one rule of the scientific method. An interesting experiment by the reader would be to look up in any dictionary the definitions of ethics and morals and the four terms still requiring definition. The definitions as given by Webster (1951) and Webster (1985) are shown in Note 1.

PART A A SCIENTIFIC BASIS OF SCIETHICS

Chapter 1 Basic Assumption and Axiom

After defining ethics, morals and laws, their basis in religion and philosophy is described. The scientific method of objective observations of Nature (reality), confirmation by others and the development of a self-consistent body of knowledge, which grows by accepting changes, additions, refinement and occasionally deletions is presented as a better process for explaining reality than supernatural phenomena.

The basic assumption for a system of ethics based on science is that:

"The most precious thing in life is life itself."

If that is accepted, then the question to be answered is, "How did we get here?" not "Why are we here?" The implicit assumption in the latter question is exposed in Chapter 2.

Following the basic assumption, we postulate the axiom:

"Those biological traits physical, emotional and mental, preferentially selected by environmental and social forces in a process called evolution, represent positive guides to acceptable actions."

The basic assumption and resulting axiom are summarized in Fig. 1.

The scientific basis is the biological and social evolution of mankind utilizing the results of molecular biology, genetics, anthropology, history, chemistry, physics and mathematics; and a new discipline, sociobiology.

The basic axiom utilized is that the exercise of the mental and physical characteristics of mankind that led to our present form are "good" or ethical because they provided the most precious thing we have, that is life itself, in our current forms. Natural forces selected these characteristics as they proved to be more effective at accommodating mankind to the environment and to subsequent changes in the environment. Societal forces have acted and are still acting to select characteristics but on a much different time scale.

1.1 Necessary Evolutionary Characteristics: The Basis of SciEthics

Any man or woman predecessor of humans must have had at least the following characteristics:

1.1.1 A will to stay alive as seen in prey fleeing predator.
1.1.2 An urge to procreate, as seen in the strong sex drive.
1.1.3 An urge to protect and nurture progeny as seen in mother love and the security of the "Marriage Contract."
1.1.4 Living in groups and tribes for the advantages of reciprocity, the strength of numbers, division of labor, specialization, leadership and the various responsibilities implied in the "Social Contract."

1.1.5 Individual freedom to choose a new course of action, which in turn provided many alternatives, which permit evolutionary selection. This is the societal equivalent of biological variability.

1.1.6 The most significant of the biological changes which occurred are:

a. The growth of intelligence, which includes curiosity and creativity. The multiple attempts to answer the crooked letter Y? led to the many concepts of God and then the practice of science.

b. The upright posture, which in turn freed the hands for carrying food, weapons and for the fabrication of tools.

c. The change from estrus to continuous sexual availability of the female, which led to the "Marriage Contract," love and the family.

1.1.7 The ability to vocalize complex information.

1.1.8 Adaptability to changes in the environment, both physical and societal.

1.1.9 The geographical diffusion of mankind, which serves as insurance against extinction by acts of nature or of man. Thus, the freedom to emigrate, "To go where the grass is greener" is SciEthical, but the right to immigrate is not included.

1.1.10 The investment of time and effort in fabricating tools, weapons, clothing, and shelter rather than searching for food or chasing females placed a high value on personal ownership of inanimate products so made.

1.1.11 With the growth of human populations, a new societal ethic has emerged, namely the preservation of the environment.

These eleven guides are summarized as SciEthics and shown in Figure 2.

Human traits are not uniform throughout mankind. Human characteristics are variable and statistically distributed. There will be many deviants from average behavior, but examples of such deviants do not invalidate statements about average behavior.

1.2 Scientific Methodology

The process of utilizing ethics based on science has three features that help in their application to specific problems. First, there is a hierarchical system of importance from 1 to 11. Second, there is a context within the science of evolution, by which the situation and consequences can be evaluated. Third, as the environment changes, these ethics, not being the unalterable words of a supernatural god, can be updated by well-established scientific procedures. In addition, there are the simple scientific rules:

a. There are no absolutes: thus, there may be exceptions.

b. The situation must be described.

c. There must be a line of reasoning which leads to a course of action compatible with the postulated SciEthics.

d. The anticipated and possible unanticipated consequences must be defined.

Examples of the use of these rules to formulate specific ethics are given later throughout the book.

1.3 Society and Science Replaces Nature's Evolution

The environment gradually changed from that made by Nature to man-made cities and streets. In addition, the societal context grew from simple groups and tribes to complex nation-states and now globalization. The non-objective, evolutionary selectivity of Nature has changed to man-made selectivity, often ill directed, destructive and ill advised.

With the understanding of DNA and the process of evolution, it is now apparent that we are beginning to be able to knowledgeably direct our own biological evolution through scientific and societal means. The basic question is do we have the wisdom to do this correctly? The time scales for these alternative methods have been markedly different. The responsibilities of society and science are described (in Part D) in the context of medical advances, disease control, genetic engineering, population density, ecology, governmental responsibilities and finally preventing Homo sapiens' extinction by foreseeable events or developments.

1.4 Flexibility of SciEthics

Not being handed down by an all-knowing god, SciEthics like any other scientific hypothesis is subject to change and correction. These changes will provide successively better approximations to reality and are not indicative of basic faults, as they would be in a religion.

Corrections, which can be made by providing more and better references to experimental results, will extend and improve the application of SciEthics. These will not take 500 years. In Part E the reader is encouraged to submit corrections and additions.

PART B ALTERNATIVE BASIS FOR ETHICS

Chapter 2 Background

2.1 Ethics and Morals

First, what are "ethics?" They are guides to any action a person takes that affects another person. This simple definition will be elaborated on later. The distinction has been made in the Introduction between ethics and morals; words which are often used interchangeably. Second, where do they come from? Usually ethics have a basis in a religion or a philosophy and sometimes in the well-established customs and practices of a culture, ie, common sense.

When someone criticizes an action you have just taken as unethical you usually argue with him or her because an unethical action is "bad." But if they praise your action as ethical, you usually agree because an ethical action is "good." The use of, and the reaction to, the words "good" and "bad," "ethical" and "unethical" (will be defined later in this chapter) is mutually understandable because your society has taught both of you the same understanding of these words.

But society always changes, new situations arise or you now think that some ethic, taught when you were young is just wrong or plain unfair. What do you now do? How do you challenge the establishment, your parents, the police, your priest or minister or rabbi and society in general? There is no universally accepted Ethical Constitution that can be interpreted by a court system to which you can turn in a dispute. Neither is there a way, universally accepted, to repeal, modify or amend an ethic as of a given date. Changes in ethics do occur, but slowly and painfully over long periods.

Where do these ethics come from and what is their basis? What makes them valid? If there are several ways to do something, what makes some ways ethical and the other ways unethical?

2.2 Sources of Ethics

There are currently four major sources or types of ethics:

Religion. A guide to human behavior established by a religion will be called a moral instead of an ethic, but the two words are often used interchangeably.

Philosophy.A guide to human behavior proposed by a mortal man, a philosopher, will be called an ethic. Often these ethics permeate the society and after many years become the ethics of the culture. Although often used interchangeably, morals usually define an approved specific course of action one is to take in a specific situation. Ethics usually provide a rational for a basic approach to take in many different situations. The word meta-ethics defines a system of ethics derived from one or a few basic principles.

Laws. Ethics that a government has codified, written and ostensibly will enforce.

Manners. Human behaviors that are well established by custom or practice but are not enforced by the government. Society, rather than government, may attempt to mildly enforce good manners by comments of disdain or severely enforce them by ostracizing the violator.

Since there are many religions and many schools of philosophy, there are many possible meta-ethical systems and moral codes. There is extensive duplication and overlap between these systems. There are also significant differences between many of these systems, which over the centuries have led to different practices often strongly in conflict.

2.3 Selection

How does one select the "correct" code of ethics or religion or philosophy? In a sense, any set of ethics defines the rules of the game of life. We can play the game as long as the rules are understood and enforced. The game might be better if the rules were different, but humanity has played this game under many sets of rules, some much worse than others.

However, the use of words like "better" or "worse" implies a ranking of rules or ethical systems, and if the act of ranking could have a scientific basis we would not be forced to make an arbitrary selection between numerous gods or philosophies. Most selections are made by default, i.e., one selects the religion or ethics practiced by their parents. As we grow up and acquire knowledge and experience, some of us change our selection of ethics and morals.

In this, book a basis of ethics other than religion or a philosophy is proposed; namely science in general and particularly the Science of Evolution.

2.4 Conflicting Claims

However, there is no way to resolve the conflicting claims of religions and philosophies. Religions that contain the statements of God obviously cannot be wrong and therefore cannot be changed even if conditions change. Philosophies developed by an individual in a certain country at a certain time may lose their relevancy with age and thus lose their believers. Modification of philosophies by later philosophers is rarely attempted.

Almost all religions and philosophies are still taught and have supporting groups. The word "almost" is used because some religions have finally died out. This usually occurs either when the adherents die, are converted, or killed. It is extremely difficult to completely remove a religion from society. Adherents of the pre-Christian religions, such as Animism in Africa and Druidism in England, are still found but no one still believes that all matter is made of air, water, fire and earth or that the whole earth is flat. (Although there is a Flat Earth Society.)

Thus, it is advanced that if a scientific basis could be found for ethics, we would be able to develop a realistic set of ethical values and have an acceptable methodology for extending or changing them when a sound basis for change arises.

2.5 A New Source

In the field of science, a process called the scientific method is used. A human develops a theory that describes a natural object or process. This theory is open to question, proved or disproved by other scientists. The basis of the theory must be described and the instruments used, if any, must be identified. Scientists in other laboratories or countries are encouraged to conduct tests to prove or disprove the theory. While there may initially be competing theories, objective physical evidence is accepted as the arbitrator that will eventually cause one of the theories to be accepted while the others are discarded. (No group now seriously claims the Earth is flat, that the sun revolves around the Earth, or that flies arise from rotting meat by spontaneous generation.) A theoretical attempt to relate religion to science is presented by Beyondism (Cattell, 1987)

The ability to accept changes in a theory is one of the powerful attributes of science. New experiments and analyses eventually resolve competing claims. As better instruments are developed and additional evidence accumulates, the scientific theories are modified or refined to accommodate and explain these new data as well, or to provide a better approximation or to extend the region of applicability. Once proven, and confirmed by independent practitioners, a scientific theory is never found to be totally inapplicable (Architects still design buildings as though the earth is flat.) but it may have to be modified or restricted in its application. However, neither is it called absolute.

Einstein's Theory of Relativity does not revoke Newton's Laws. It merely restricts their area of applicability to the macroscopic world and to speeds substantially less than the velocity of light. Engineering today is 99.9% still based on Newton's Laws, the results of which provides an excellent approximation to everyday physical phenomena. Under unusual circumstances, Einstein's Laws provide a better approximation. But no scientist will claim that Einstein's Laws as now defined provide the last word in the field of physics. All accept the possibility that someone will come along with a new formulation that will provide a closer approximation to reality over a wider range of conditions.

The structure of scientific knowledge thus grows to explain our world or Nature and finally ourselves without conflicting explanations except temporarily at the "cutting edge" of new findings and new theories. The house of science grows somewhat in the way a coral reef grows, by the accretion of bits of new data, refined definitions, better approximations and occasionally a new branch growing in a different direction. Occasionally a branch breaks off.

In a sense, science is never wrong, it is just an inaccurate approximation. New findings, like Einstein's Law of Relativity supplies an explanation of Nature that is more accurate than Newton's on the atomic scale and at high velocities. Newton's Laws are not wrong unless one insists on absolutes, which is a fallacious religious dogma.

2.6 The Science of Evolution

The scientific basis presented here is the biological and social evolution of mankind. The formal fields are biology especially molecular biology, sociology, genetics, anthropology, chemistry, physics, mathematics; all now integrated in a new discipline called sociobiology. (Wilson 1975)

Social evolution is considered to now play a very strong role, being a result of and in turn causing biological evolution. Selection from biological variants can be made by social forces as well as by environmental forces. These social forces predate Homo sapiens. The basic concept utilized is that the mental and physical characteristics of mankind evolved to their present form because certain differences proved to be more effective at accommodating hominids and later mankind to the then existing environment and to the changes in the environment and were thus preferentially selected. Thus, the axiom detailed in Figure 1 is developed as follows:

For any man or woman (or any animal or living thing), the predecessor of humans must have had at least the four following (necessary and sufficient) characteristics:

A will to stay alive. The desire to stay alive is evident in the struggles of a drowning person, the frantic speed of a prey fleeing a predator and the twisted tree growing out of a small crack in a cliff. The determined efforts of survivors of a sunken ship to stay alive on a raft waiting for rescue while enduring incredible hardships is the classical example. Snowbound stranded travelers have resorted to cannibalism. The desire of prisoners and slaves to stay alive although their social environment may be very stressful is another sample of the will to live, a genetic trait. (The question of suicidal actions will be discussed later in Chapter 8.)

A strong desire to reproduce. Once successful in staying alive, any living specimen must leave offspring or we would only see that specie in a museum. Again, in order to propagate the specie, there must be a strong urge to copulate, which we now call the sex drive. In this respect, there has been a remarkable over kill in Homo sapiens. Although sex from 20 to 40 years of age would currently be adequate for propagating the specie, the sex urge is strong from 13 to 75, and, for females, long after conception is possible. The urge in the early years, i.e., 13 to 20, may reflect that reproduction as soon as possible had a higher probability of success when the life span was much shorter in pre-historic times. It is difficult for a man over 60 to remember the hornyness, the constant erection and the extreme sex desire of a 17-year-old boy. (I can't speak for the female sex.)

A desire to protect and nurture. Pertains toward the resultant progeny until it was self-supporting and able to continue the two preceding characteristics. Giving birth is in of itself a necessary but not sufficient condition for propagation of the specie. Human offspring must be fed, nurtured, taught skills and protected until they are capable of self-support and procreation. While many animals may be able to move around and to find their own food in a few days or weeks, the human baby is now not capable of finding his own food for at least 15 years. Thus, the nurturing and protection of the young is another necessary condition.

A sense of responsibility. Directed to the group or kin lived with, to protect and promote their well being as well as his/her own. This sense is usually exhibited in altruistic behavior to others, such as sharing food, protecting against harm and helping in the upbringing of the progeny of others. Such action is in reality selfish as it increases the probability of success in propagating your own genes through two mechanisms. The first is the obvious advantage of reciprocity. You or your progeny may be the ones who are helped in times of emergency. The second less obvious mechanism is the perpetuation of your genes via siblings, nieces and nephews. Although there are some animal species that do not socialize, most cooperate and live in groups. (Dugatkin, 1999). The tremendous progress of Homo sapiens and other mammals that live in groups would not have been possible without cooperative group activity.

If these four basic evolutionary characteristics did not exist in the human race, we would not be here and any concern about ethics would not exist. Thus, the basic assumption of scientific ethics, namely that the most precious thing in life is life itself, leads to the axiom:

Biological traits—physical, emotional and mental—preferentially selected by environmental and social forces in a process called evolution, represent positive guides to actions that will perpetuate the specie.

These guides are considered to be the basis of scientific ethics (SciEthics). See Fig. 2 for the detailed list of SciEthics. These detailed guides will be described in greater detail in Part C.

2.7 Why Are We Here?

In science there is a saying, "Asking the right question is 50% of the solution to any problem." The question often asked, "Why are we here?" is the wrong formulation of the question. Why are we here is a question that contains an assumption. A better-known example of this logical fallacy is, "Why are you still beating your wife?" The question assumes a condition that may or may not be true. If it is not true there is no way of answering except by a denial of the assumption. "Why are we here?" has the hidden assumption that a god, or something, had a reason for putting us here. If there was no one then the question cannot be answered in a meaningful way. The more useful question is, "How did we get here?" which would lead to the explanation of, "Why we are here." Note that this formulation does not exclude the possibility that a god or some one or some thing put us here for a reason we cannot fathom at present. The question of what was there before the Big Bang and what started the Big Bang is currently not answerable. The development of an answer based on science may possibly reveal that there is a God. Thus, to be conservative, one should be an agnostic not an atheist.

2.8 Definitions

Many words have more than one meaning. Usually one can select the correct meaning from the context. But not always! As part of the scientific method, critical words are explicitly defined. It is hoped that the glossary and definitions provided herein will minimize misinterpretations.

2.8.1 Words Are Labels

Words are primarily labels for something else. The something else may be:

Objects. Things that can be seen, smelled, touched, heard or tasted, that are detectable to our five senses are called objects. Technology may extend the range of a sense so that we can, in effect, see things that we normally cannot see. X-rays and MRIs represent extensions of our ability to see. Audio amplifiers represent an extension to our ability to hear low sounds and very high frequencies. There are technical amplifiers for all our senses.

Physical Processes. When objects physically interact with and affect one another, and/or form a new object, it is called a physical process. Sometimes a specific object changes, such as when a person ages, without any apparent interaction with other objects. However, the environment and internal biological objects called genes usually cause these changes. Thus, ice apparently melting by itself, or a person aging are also processes.

Mental Processes. Ideas, emotions and desires that are the result of complicated physical processes between different parts of the human mind are called a mental process.

This could be either the conscious or the unconscious mind. The results of this process can be kept in the mind in secrecy from the conscious mind.

Behavior. Physical acts by humans that reveal to others ideas, emotions or desires by movements of the body or parts thereof to effect a desired result is called behavior. This includes speech, body English and physical force.

Words are extremely efficient tools for us to communicate with each other. Try describing an eclipse without using the words "sun" and "moon." Words are labels for the objects, processes, ideas and feelings they refer to. However, words have acquired characteristics of their own in addition to being merely labels.

The Evolution of Words.

With the growth of language, new meanings are ascribed to old words and new words are invented outright. Lazy authors use familiar words in a new sense, which is a hazard to the reader who thinks of the old meaning. More innovative authors compound new words from parts of old ones, from dead classical languages or even from obscure current languages. While this leads to less confusion in one sense it has a danger for the reader with a short memory. Although a glossary is often provided, too many new words create a text that reads like a foreign language. This is a common problem for someone reading a highly technical or legal text filled with many unfamiliar words.

Thus for scientific or pedagogical clarity, key words should always be defined in the sense that the author intends its use, if that differs from the standard dictionary definition or from the prevalent common usage. This is of extreme importance when presenting a new theory or concept. Really new words are often required.

2.8.2 Pejorative Words

With usage, some words become endowed with characteristics other than those of simple labels. Motherhood, love, apple pie, God and country, dirt, kindness, compassion, fairness, alien, vile, shitty, etc. are all able to arouse the emotion associated with the referent even if the referent is not involved. The label has become a substitute for the emotion associated with the referent and signals one's attitude, opinion, judgment and final decision even before the facts can be presented, considered and judged. These words, called pejorative, essentially indicate a prejudgment opinion and their use should arouse one's suspicion.

In order to avoid the emotional reaction to the use of these words, scientists instead use cool and detached words that are not pejorative, but act as simple descriptive labels. Politicians running for office do just the opposite. Preachers from the pulpit do likewise.

2.8.3 Judgmental Words

There is much confusion about the words good, bad, wrong, right, and others that are judgmental in nature. For the reasons given above, these words are now defined as they will be used in this book.

Good /Bad/Neutral. The judgment that an object or process promotes/ hinders/does not affect mankind's ability to adapt to the physical and/or societal environment.

Right/Wrong/Neutral. The judgment that a human's behavior enhances/diminishes/does not affect mankind's ability to understand the physical and/or societal environment.

True/False and Correct/Incorrect. The judgment on a statement that can be proved/disproved by scientific methods.

Uncertain/Nonsense. Judgment on a statement that cannot be proven at this time by scientific methods because the tools are not available, i.e., telepathy is uncertain. A perpetual-motion machine is nonsense because it will never be available as it is in absolute conflict with the precepts of science.

Moral/Immoral/Amoral. Judgment on an action by a human that is approved/disapproved/without moral content based on religious tenets given by a supernatural being and as interpreted by a human prophet or scholar.

Ethical/Unethical/Neutral. Judgment on an action by a human that is approved/not approved/ without ethical content based on philosophical tenets established by a human scholar.

SciEthical/UnSciEthical/Neutral. Judgment on an action by a human that helps/hinders/does not affect a person's ability to follow the Primordial Laws (PLs) of SciEthics.

2.8.4 Definitions of Words Used in This Book

Moral: A rule for human behavior based on a specific religion. Other religions may use the same moral or a different one to define the acceptable human response to the same situation.

Ethic: A rule for human behavior in a given culture about a given situation usually based on a philosophy or custom.

SciEthic: An ethic based on science, as developed herein. There may be duplications and differences with morals or ethics.

Mankind: This includes men, women, children, the five races and any member of the species Homo sapiens. Technically, if the mating of a male and female results in a fertile offspring, the male and female are members of the same specie.

Hurt: Physical damage to the person.

Harm: Identifies damage to objects and/or finances.

Offend: Identifies a violation of concepts, morals, ethics and feelings. It does not include physical or fiscal damage to the individual.

PART B

Chapter 3 Religion

The basis of a religion is usually a supernatural event or Being which started the world. The most familiar one to the Western World is the One God who created the world and all the inhabitants thereof in six days and rested on the seventh. From this, we inferred that we should work six days and rest on the seventh. This is an example of the conversion of a facet of our societal growth into an ethic. This practice of one day of rest in a seven-day week is thus a moral from a religious basis.

However, the One God had three prophets with three different messages so the day of rest can be Friday, Saturday or Sunday depending on whether you are a Muhammadan, Hebrew, or Christian. If adherents of two or more of these religions inhabit the same area and if the members of the majority religion take the choice of the specific day of rest seriously and try to impose that day on adherents of the other religion, we have the start of religious warfare. All because of different moral values arising from the same religious beliefs.

It is unfortunate that the One God had three different disciples; Moses, Jesus (claimed to be the Son of God by most Christians.) and Mohammed. They provided many different interpretations of His Word. This simply reflects the fact that three charismatic individuals, leaders of their time, saw things differently. Followers of the three usually do not like the presence of people who disagree with their specific interpretations. The concept of a supernatural person with infinite wisdom who created everything does not permit the presence of people who disagree with any specific wisdom because this throws doubt on their basic concept. The fact that some members of the tribe in which Jesus was born did not believe in his divinity at the time of his existence is very disturbing to the followers of Jesus. They become anti-Semites and try to convert or exterminate the descendants of the tribe that to this day still refuse to accept the divinity of Jesus. This illustrates the danger of fanatical fundamentalist's belief in unchangeable, engraved in stone, religious morals in an admittedly simplistic manner. However, how an anti-Semitic religious Christian can worship a Jew is puzzling.

Scientists on the other hand, will somewhat slowly accept new evidence and change or modify or reinterpret theories or accept boundaries to the theories.

3.1 The Ten Commandments

The Ten Commandments of the Bible, as listed in Figure 3, are accepted by all the Judo-Christian sects as God-given, that is provided by a supernatural being to a human, Moses, engraved on tablets of stone.

An alternate interpretation from the scientific basis of ethics is now postulated. Moses was a leader of a diverse group of ex-slaves unused to self-government and with many old practices not relevant to their new life in the desert. He realized there had to be an acceptance of a common set of realistic ground rules that would integrate these unruly people into a coherent cooperative tribe. The many miracles "performed" by or attributed to their God that effected their escape from Egypt did not provide any set of laws that would govern their behavior. Although they believed in their One God, this did not prevent them from creating a golden image to worship. Moses in addition to being a great leader was also a superlative social engineer. He knew that to make them into a one people they would have to have many things in common by which to identify themselves.

Commandments One, Two and Three

One of the common elements Moses identified was their One God who also had provided the "miracles." Thus, the first three laws established the supremacy and exclusiveness of the One God.

Commandment Four

Moses had probably observed the improved productivity of slaves after a day of rest due to rain or a holiday and made the insightful extension of this rest period to a periodic full day. The selection of 7 was probably due to some myth of the time. This periodic day of rest, independent of national holidays, an unusual feature then, would also provide another unique feature of tribal commonalty. Thus, he made the Fourth Commandment to observe the Sabbath.

Commandment Five

In the societal development from groups to tribes to nations, particularly before written records, much of the history, experiences and knowledge were in the memories of the older members of the group. Since they could no longer hunt, forage or fight, their support represented an apparent economic burden. Yet, with the benefit of insight, it was apparent that their presence provided a source of history that was an asset, which overweighed their liability. Also they could baby sit and help in the nurturing and education of the young. They also constituted an insurance policy in the event the parents died for any reason. Thus Moses provided Commandment Five, namely, honor thy father and thy mother.

Commandment Six

He included the First Primordial Law by Commandment Six, forbidding murder.

Commandment Seven

The practice of monogyny is reflected in Commandment Seven, which forbids adultery. This is apparently counter to PL No. 2, which approves of men and women leaving as many offspring with their respective genes as possible. It may be considered an inadequate version of the Primordial Law No. 3 that states you must rear your young until they are capable of supporting themselves. In adultery, a man may not be available to support all his possible offspring. However, an alternate statement such as, "Don't have more children than you can support and don't steal your neighbor's right to pass on his genes" reflects many other SciEthical rules to be considered later.

3.2 Morals

The Ten Commandments illustrate the development of morals by religion. There are many more morals available via the Sermon On The Mount (Christian) (See Note 9), the Koran (Muslim), the Bhagavad Gita (Hindu) and many others.

12

As this book is focused on the ethics developed by genetics (Nature) and by society (Nurture) rather than on religions (and philosophies), no attempt will be made to present all the morals available from the many religions nor the conflicts inherent between them. That would be a book by itself. The Ten Commandments have been used just as an example of morals established by religion. Ethics established by philosophies will be covered in the next chapter.

3.3 Problems With Religions

3.3.1 Inflexibility

The Hebrews have often, tortuously, reinterpreted their Torah as conditions changed. The Catholic Popes have changed their interpretations by issuing Papal Bulls and Encyclical letters. However, rarely have they changed any dictum that was clearly stated in their "Bible." There are a plethora of sacred texts of religions, and there are many sects in which the written word is taken literally as the "Word of God" and cannot be changed. This is especially true of the followers of Mohammed who oppose any change in the Koran or in its interpretation. This attitude of the "fundamentalist" sects of all religion is not conducive to the evolutionary process of adaptation to changes in the environment.

3.3.2 Unresolvable Conflicts

The three major western religions as presented by three different prophets have differences. But since the words of God are inviolate, there is no way to resolve the conflicts that arise from these differences and from the different practices that have developed with time. Thus, the proponents of one religion will fight with the proponents of a different religion and kill them although they both worship the same god. The zealots of one religion cannot accept the existence of large groups that have a different religion and live peaceably side by side with them. A rough peace has developed in modern times but there are still deadly conflicts. Northern Ireland, Sri Lanka, Yugoslavia and Israel are only a few examples. Invariably there are economic and territorial differences that are aggravated by religious differences. There always have been purely religious conflicts such as the attempts to reclaim the lands of their religion founders by the Crusaders and the Zionists. Genocides also still persist although they are territorial and economically motivated, aggravated by religious and ethnic differences. Germans/Jews, China/Tibet and Soviet Union in Latvia, Lithuania, Estonia. Turkey with the Armenians and Kurds is another example.

3.3.3 Discredited Hypothesis

At the time of the beginning of religions there was no objective evidence or facts to answer the many questions of WHY nature was the way it was or WHO started all this and WHY and HOW? With the strong feeling of curiosity inherent in the growth of intelligence, answers were required. The wise men of the tribe invented stories of a supernatural being, a god who started the world. There are as many different stories, different ways the different gods created the world, and WHY as there were different tribes and different wise men. The one most familiar to us is the story in Genesis of how God created the world in six days and rested on the seventh. God created each of the species of plants, animals, birds, and insects that were evident. Also, that man was made in God's image.

For many centuries, this explanation held up, as there was no basis to question it. With the development of the science of archeology, anthropology and the ability to date past events, it soon became possible to question whether the world was created some five thousand years ago. Darwin's Theory of Evolution furnished a better explanation of why there were so many species. Radioactive dating of different elements provided strong evidence that the world was millions of years old. Thus, the hypothesis of supernatural gods creating the world fairly recently, which is the basis of religion, has become discredited. Very few educated adults really believe the story in Genesis literally.

Many attempts have been made to justify the story of Genesis by calling it allegorical rather than literal; by noting that the sequence of events is roughly scientifically correct but on a different time scale; and by questioning the accuracy of the written story which had originally been handed down verbally over many generations. These efforts are not very successful. One must have absolute faith to believe any of these genesis stories literally and to claim that when God created the world, he also created all the evidence that has been found that questions the story. If man is made in God's image, then he must have many different appearances; white, brown and yellow. This is difficult to understand. Thus, the hypothesis that all the various aspects of nature can be explained by the existence of a supernatural and all-powerful god is now discredited. This discreditation also casts doubts on the societal laws based on God's words. In effect, science has discredited the notion of gods creating the world as described in the various accounts of the origin and nature of the world.

However, there are many scientists who, not knowing what there was before the Big Bang, or what caused it, prefer to call themselves agnostics rather than atheists. Until more evidence is in, they are reluctant to say there is no God.

3.4 Reasons For The Popularity Of Religion

3.4.1 Introduction

In the prior section on the basis of ethics, a strong distinction was made between religion and science. Explicit statements were made that showed religion had several fatal weaknesses, one, the belief in supernatural phenomena, and second, the difficulty to change when conditions changed. A basic lesson from science is that there are no supernatural phenomena (only phenomena we do not yet understand), and from evolution it is that a specie must change when the environment changes markedly or it will become extinct. As the environment today is mostly societal in which religions play a large part, it may appear strange that religions have remained strong and persistent in spite of the fact that science has really eroded their historical basis. How is this explained? If religion is so wrong technically, why has it not become extinct? A personal explanation follows.

It appears that there is a scientific basis for the strength of religious beliefs. There is also the realization that they served a useful purpose in the evolution of mankind although it is believed that many religions are now detrimental to that end. The scientific basis of religions and their usefulness are now described.

3.4.2 The SciEthic Basis of Religion

The basic assumption of SciEthics is recapitulated in Fig, 1. If you agree with the basic premise that the most important thing in human life is life itself, than it is difficult to fault the logic that leads to the eleven SciEthics listed in Fig. 2. The positive values of religion were unknowingly based on some of these SciEthics and the scientific theory of evolution.

It is postulated that during the evolution of mankind, the most important biological change was the increase in brain size and intellectual capacity. Thus, PL 6 provides the guide that we should value intelligence above physical strength and speed. The former allowed us to defeat the larger, faster and more ferocious mammals that were part of the environment.

Long ago and far away, somewhere on the path of evolution, after the ability to speak in abstract terms had developed, people started to ask WHY. WHY the sun moved across the sky in the daytime? Why the moon did the same at night? What were the stars? Why there were storms, lightning, rain, so many different animals and fishes and birds and trees, and why did my lovely little child have to get hot and die?

As part of the growth of intelligence, individuals developed a curiosity of WHY Nature was the way it was. They also developed a creativity that resulted in useful answers as well as artistic items that were selected in the societal evolutionary process. Thus, curiosity became a favored genetic trait. This curiosity is evident in small children who ask thousands of questions, some so penetrating the parents often are unable to answer. Often the questions pertain to myths and have no logical answer. We just laugh them off or try to explain the lack of reality of the context of the question. Sometimes we get so exasperated at the number and nature of the questions that we are forced to use the universal response, "SHUT UP." The reason to dwell on this childish trait is that the characteristics exhibited by small children such as, fear of falling, fear of the dark, putting everything they find in their mouth, and the asking of many questions is believed to be genetically prompted and reflect the evolutionary path of development. These children are too young to have been influenced by education so that their behavior is due to nature and not nurture. Thus, the asking of questions is a basic property of mankind, and it is the creative attempts to answer them that has led to most or all the improvements made by mankind.

Curiosity and creativity are good traits and should be encouraged at all ages. Any adult efforts to quell curiosity or reject a creative effort out of hand are unSciEthical, The use of myths such as Santa Claus, the Tooth Fairy and the Stork is a disservice to their education. Unless of course, one considers that the child's disappointment when the truth is learned is a lesson not to believe everything you are told. True explanations that can act in the same fashion as these myths should be formulated and used.

In attempts to answer all these questions and many more, man created the concept of supernatural beings that did those things. At first, there were many supernatural beings. As an example, there was Jupiter, the major King of Heaven; Juno, his wife who portrayed the Earth and fertility; and following the known pattern on Earth, they had many children in different ways. One for each phenomena. The big advantage of this personification of natural events was that a supplicant could appeal to someone to arrange things to avoid the disasters that could occur or to provide rain or a good haul of fish. Perhaps a gift to the gods would help?

The Wise Man of the tribe, acted as a caller on the gods and as an interpreter of his response or lack of response. This is an example of the specialization of services that became possible as the size and efficiency of the tribe increased. Depending on the circumstances, one might have to appeal to different gods. Thus, Animism as a religion provided answers to the many questions that arose. By statistical chance, it often appeared that a gift was adequate and the plea had been answered positively. If the answer was negative, the Wise Man could suggest that the next time maybe the gift should be larger. At least there apparently was something you could do to influence the forces of Nature.

Because of the continued growth of intelligence and knowledge, it slowly became apparent that these personified gods really did not help. Sometimes you were lucky and obtained the gift of the gods without an appeal or sacrifice. Gradually the many personified gods were abandoned for an ethereal god who never was visible. This is the One God of Western Civilization who gave us Judaism, Christianity (although the Son of God and the Holy Ghost were a throwback to the Greek and Roman mythologies) and finally Islam or Mohammedism. This One God had the power "To move in mysterious ways His wonders to perform." Thus, anything that happened could be explained away. This is certainly a logically neat answer to the many questions asked of the Wise Men.

3.4.3 The Utility of the God Concept

The leaders and prophets, who were men of above average intelligence, created socially useful rules and laws. They then used the claim of God's divine revelations to obtain support for these rules and laws that were beneficial to the individuals in a tribe and essential for them to operate successfully as a group. This is illustrated in the earlier discussion on the Ten Commandments. Most of the rules were sound and we now note that many of them are equivalent to some of the SciEthics shown in Figure 2. These early rules and laws were appropriate to the then current environment.

However, as the environment changed with time, these early rules and laws had to be changed or re-interpreted in order to be applicable under the new conditions or under a more scientific understanding of nature and reality. However, these rules had been endowed with the authority of a supernatural god who could not be wrong. Changing them became very difficult because it would admit the potential fallibility of God's way. If he could be wrong in one case, why not in another? The way out of this bind is to reinterpret the law. This is often done. Sometimes it takes 500 years for this to happen (Catholic Church re: Copernicus). Religion does not have the flexibility of science to change its interpretation of nature upon finding new objective evidence.

3.4.4 Sociality

Another advantage of organized religion is that the rituals of observance in a house of worship bring people together and provides an opportunity for communal bonding. It provides a structure in our lives of what to do and when to do it. A sense of continuity develops as we go to church every Sunday. It provides an activity that binds the family as a whole and then with friends that are made during the non-service activities such as potluck dinners, fund raising and facility maintenance. You become an accepted member of a community. Your children make friends. These acts strike a deep chord in our emotions placed there by the advantages of tribal living selected by evolution. This is reflected in PL 4.

3.4.5 Entertainment

Other advantages of organized religion is that for many people if provides an escape from their humdrum lives. This is through fantasy or ecstasy as illustrated by the speaking in tongues, healing by the lying on of hands, and occasionally it provides orgasms to virgins. The pomp and ceremony of religious holidays reinforces the sense of tribalism. The singing and chanting together strengthens the sense of brotherhood and sisterhood.

3.4.6 Gender Support

Since men and women are different in many ways beyond obvious physical structure, the church affairs provide an opportunity for the women to talk to other women, to exchange information on services, shops and products, i.e., "woman's talk," compare husband behavior and get help without the inhibiting presence of their men. Similarly, the men have an opportunity to do the same thing with other men without the inhibiting presence of their wives. This is not to deny that there are other groups that offer the same benefits, but the church has become very convenient. It is a successful and profitable social organization.

3.4.7 Mobility

When people vote with their feet in favor of a distant country, they lose their societal base. In a high tech, always rapidly changing economy, there is need for workers to be readily mobile and to move around the country. In addition, our aging population tends to move from cold weather climes to more pleasant weather in Florida and the Southwest. The ability of these immigrants, workers and "snow birds" to rapidly become accepted members of a new community is another positive value of organized religion. One has to "belong" to a small intimate group as well as a large nation. There is little intimacy between an individual and 250 million fellow citizens.

3.4.8 Purpose Of Life

Perhaps the deepest felt value of religion is its providing answers to the questions "Why am I here?" or "What is the purpose of life?" These questions reflect Homo sapien's high intelligence, which is trying to understand everything about everything. The fact that religious based answers to these questions are imaginative but false really does not matter to most people. These answers are widely believed, so why not believe them too? We are not all intellectual giants. Man has lived for many centuries with false, i.e., incorrect answers to many questions. (It does not make much difference to the bulk of the population whether the sun goes around the Earth or vice versa. But it makes an enormous difference to the growth of knowledge.)

By telling you you were made in God's image your self-respect and self-worth is enhanced regardless of your worldly status. By promising the potential of life eternal in heaven, you are provided with a purpose in life, namely to follow his precepts so as to attain that goal. These assurances may satisfy you and make a hard life bearable. If you believe that you are made in God's image, then that cannot be improved and there is no point to spend any effort in improving yourself. Then if by following his rules you will have life eternal in heaven after you die, there is little point to struggle to improve your current life. These two religious claims are false and basically harmful. They make you accept all the injustices and exploitation that lead to a miserable life in the here and now which is the only life you will ever have. They help the individuals in power to maintain their dominance even if most of the people are poor, miserable and desperate. This is a con job par excellence.

There is no a priori reason for Homo sapiens to exist in the sense that some intelligence placed him there for a reason. In that sense there is no purpose to life. However, since the most precious thing in life is life itself, we can say that the result of all our ancestor's activity was to produce us, then that was their purpose albeit not consciously. In this sense, the purpose of life is to perpetuate the specie and improve its probability of continuation by increasing its intelligence and power to control its environment. In this sense, the purpose of life is to have progeny and make their life better than our own. This can be done in many ways, from improving their individual genetic composition to improving their societal environment. Thus, the purpose of life is to perpetuate life and make it better.

This argument has a circular ring to it, yet the logic seems sound. If I look backward to my ancestors, their actions were aimed at making a me and making my life better than theirs. This intent has a familiar ring to it. All parents say they want their children not to have to go through what they went through, to be better educated and better off then they were and thus have a better life. The objectives stated are usually materialistic, but often unstated are silent promises to spend more time with their young children than their parents had spent with them, devote more time and effort to family affairs than to business or to the job or to golf. This is the attitude noted in most poor and middle-class people of most ethnic groups and faiths.

It is only with the development of science that answers capable of being demonstrated, observed and replicated by others, have become available. Thus, we now know that the correct question is "How Am I Here?" not "Why Am I Here?" Of course, a major advantage of correct scientific answers is that it serves as a foundation that permits the addition of new scientific facts and thus the continued growth of knowledge of the universe we live in.

The scientific answers developed by humans do not answer all questions at any given time as is done by proponents of religion. The Koran, written about 500 AD, provides all the answers needed by an observant Muslim 1,500 years later. A strict adherence to this belief will ensure the stultification of the Islamic culture or civilization.

As always, scientific information can be used for good or evil. The application of this information generally falls on the government.

There is a conflict between people wanting to use the data or drug that holds some promise because their loved one is dying, and the professionals who do not yet know what the side effects may be. When such data first becomes available, the government's problem becomes very difficult because early data are not black and white, yes or no, but usually probabilistic in nature, greatly extrapolated and subject to questions arising from incompleteness. However, an eventual answer to these questions will be provided by science (It will not be provided by Super Mom.) and the drug or procedure will eventually be safely available. However, all of this is poor comfort to someone whose loved child or parent is dying right now.

3.4.9 Immortality

Evolution led to the love of life (PL 1) and the fear of death. These traits of human feelings and behavior were selected by Nature because you could leave progeny only if you lived long enough to copulate with a spouse and provide food, protection, training, education and love to the offspring for 15 to 20 years generally; maybe 30 years nowadays if you wanted your son to be a doctor.

Thus, these behavior patterns were favored and passed on. Wonderful you may say, but then the great dichotomy develops. Having bred in you this love of life, Nature takes it away by also making you mortal. Eventually you die. We all die. This is a terrible contradiction. You get the gift of life and then it is taken away. Science can only say, too bad, but that's the way it is.

Then you are offered a solution to this awful dichotomy. Believe in a god, follow his orders/rules as a man in a black suit interprets them and after you die in this world, this god will raise you to his world called Heaven and you will live happily ever after. Boy, I'll buy that! First, what are the rules? Oh, follow the Ten Commandments and the Golden Rule. Hey, that's not too bad! They make sense. I even like most of them. Sure, sprinkle holy water on me and sign me up. Thus, it turns out that a religion offers a solution to one of many problems that evolution has created. For example, mother love is an extremely strong emotion that was selected by evolution because it increased the probability that in times of stress children would be protected, fed and nurtured, and thus survive. Yet when a child does die, science just says, tough luck—that's the way the cookie crumbles. Try again. If the child is 25 and you are 45, that's rather difficult advice to follow. It is disheartening to lose a child. Many years must pass before a bereaved parent can operate somewhat normally again. And she/he never really forgets.

Religion has stepped in and provides a promise that you will see your child, your spouse, and all your loved ones again in Heaven after you die. So say your prayers, be comforted and all will eventually be well. That is a solution that is heart warming and difficult to reject.

Thus, the ability of religion to answer some of the difficult questions that arise in life contributes to its popularity and power in society. The answers are very satisfying to most people, although they are always of a structure that cannot be proved or disproved in a scientific sense.

The scientific answers developed by humans do not answer all questions at any given time as is done by proponents of religion. Can one believe that the Koran, written about 500 AD, provides all the answers needed by an observant Muslim 1,500 years later? A strict adherence to this belief will ensure the stultification of the Islamic culture or civilization. The ability of science to answer questions grows slowly like a coral reef, with small bits of data that add up. To expect immediate answers to all questions reflects the mind-set of a religious person who believes in an all-knowing God.

3.4.10 Conclusion

Thus, we must concede the many positive values of many of the morals established by religion, the support of societal (tribal) operations by rituals and practices, and finally the peace of mind offered by religious answers (although they may be wrong) to the questions raised by man's deep seated genetic desire for answers to WHY?

As adults we still ask questions like, "Why are we here? What is the purpose of life?" The answers provided by science are usually not emotionally satisfactory except perhaps to those who get a deep feeling of satisfaction from the acquisition of knowledge.

3.5 Atheism

3.5.1 The Atheist

What does the atheist offer the general public for emotional satisfaction? The atheists who focus on the absurdities and errors in religious thinking may be correct in a narrow technical sense. A mythical god did not create the world in six days and rest on the seventh. When you die, there is no heaven to receive you even if you have obeyed certain rules and made certain sacrifices, which are different for different religions. Neither God nor his offspring keeps track of each individual's activity so that you have to settle your accounts with him before he will bless you.

All these are the absurdities of an anthropomorphic deity invented long ago by rational and very astute men in every society. These men however did not have the benefit of 20th century science.

By focusing on and fighting these absurdities, atheists deny the positive values and functions of religion which make it attractive to so many people all over the world. Atheism is a very cold philosophical belief that has never gained a large following. More seriously, it is a negative emphasis on the faults and fallacies of established religions and does not seem to offer any alternatives to replace religion's positive values

3.5.2 The Agnostic

If you believe, positively, that gods do not exist then you are an atheist.

If you doubt that gods exist, then you are an agnostic. This position is somewhat supported by the fact that science cannot prove or disprove the existence of God. What caused the Big Bang and what was there before the Big Bang are unanswered questions that cosmologists are still struggling with. But that uncertainty does not affect the postulations used in this theory of the scientific basis of ethics. We have shown why belief in gods is so popular; now we will look at why Atheism and Agnosticism have so few followers.

3.5.3 Lack of Emotional Appeal

The concept of any atheistic philosophy, as well as Scientific Ethics, lacks an emotional appeal to the average person. There is no one to turn to in times of trouble, no one to ask for forgiveness when one has done wrong, no one to blame when something goes awry (no "It's God's Will"), no one to explain away incredible injustices (no "God moves in mysteries ways His wonders to perform"), no one to provide an answer for why we are here, etc. The appeal of a personified god who can explain all of the above goes very deep into a person's psyche, and for a very good reason.

Many of us, when we were small children were afraid of the dark. We did not know what unknown dangers or monsters lurked in the closet behind that closed door. We asked our mother to leave a light on or keep the bedroom door open. This phase usually did not last long but it reflected a deep genetic lesson learned millions of years ago; be careful of the dark or any place where you cannot see into. There really may be dangers there, dangers from wild nocturnal hunting animals. The early fear of the dark was well founded and is today still applicable in many cases. Its existence in small children indicates it is genetically derived because the small child has had no experience to teach it that lesson. (This also indicates that fire must have been invented/discovered quite recently in the evolutionary sense because the small child has no fear of the fire or the glowing embers. It has to be taught that fear by letting it touch hot but not damaging items while we condition it by saying, hot! hot!. Nurture, not Nature is the active force here.)

When the small child cries out in fear, whether this is due to the dark, a nightmare, or just being alone, mother comes rushing in and comforts the child. Being held in strong arms against a warm bosom and with the familiar smell of its mom, the child is soon soothed and falls asleep again. Mom has vanquished the hazards of the unknown goblins in the dark closet. As the child grows up, it usually loses its fear of the dark as a specific hazard. But as the child grows up it discovers other unknowns, events it could not predict, actions by others that are threatening and darknesses that are psychological, not physical. Mom is no longer able to comfort the growing or grown child or adult. More realistically, the grown child is no longer able to believe and accept the mother's reassurances.

However, society has supplied him with a Super Mom. Someone that does have all the answers, can comfort him in his problems and promise him ice cream later or eventually paradise and immortality. This Super Mom is God who can explain everything, promise anything and who does provide comfort in times of bereavement. The adult transfers, in his psyche, his childish feelings for his mom to his God. This does psychologically comfort him. No question of it. People are comforted by the thought that God will take care of things or will forgive you for wrongs you have done.

3.5.4 Only Probabilistic Answers

Scientific ethics or Humanism can only offer a probabilistic answer to many of life's tragedies. For example, people at every level of society, rich or poor, educated or not, have a risk of developing cancer. The risk is quantified and expressed in numbers as a percent or cases per million. These numbers reflect reality but it does not satisfy the average individual. It is cold comfort, not the same warmth as your mother's arms or unquestioning belief in her reassurances when you were small.

Although science is currently unable to be specific on what went wrong, or what could have been done to avoid a cancer, science is slowly gaining the knowledge to answer those questions. The Genome Project will provide the information on the genetic contribution to various maladies. New equipment will diagnose cancer earlier, making it easier to treat successfully. Testing on animals has provided some and will provide more information on specific drugs that can combat cancer and on specific chemicals used by industry, that can cause cancer. All of this is poor comfort to someone whose loved child or parent is dying from cancer right now.

However, an eventual answer, prevention, and remedy will probably be provided by science. It will not be provided by Super Mom. The only satisfaction people dying from a disease have today is knowing that their children, grandchildren or maybe their great-grandchildren will probably not die from this specific disease if scientific investigations are continued. Today we have little fear of dying from leprosy, chicken pox, diphtheria, polio, and countless other diseases that plagued our parents. Science holds the promise for real solutions, but this is poor comfort for today's tragedy. If you want this comfort, go ahead and commune with the Super Mom but keep science supported. It's a little like the WWII song, "Praise the Lord but Pass the Ammunition."

PART B

Chapter 4 Philosophy

We now consider the other basis of ethics, namely philosophy. In this section, there is no intent to present a thorough analysis of philosophy. This author is in no way qualified to do that. Instead, the highlights of what are the two main schools of Western philosophy believed to be pertinent to SciEthics are very briefly described.

4.1 The Greeks

Modern philosophy started in Western culture at about 500 BC when Pythagoras started using logic and mathematics to explain reality. He was followed by Socrates who defined ethics based on logic and not religion. His efforts are reflected in Plato's Republic, which described an ideal society and the code of behavior of its inhabitants. This of course was ideal in Plato's sense of what was ideal. Slavery was acceptable. There is no intent to deprecate Socrates or Plato or Epicurus who followed. They were intellectual giants for which there were no equals for about two millennia. The point that needs to be made is that the ethics that flowed from their philosophy were based on human thoughts and did not have a supernatural origin. This was a tremendous step forward. (Tarnas, 1991) A comparison of early Greek ethics and SciEthics would be interesting.

4.2 The Enlightenment

Philosophers in the mid-seventeen century started to analyze human behavior from a scientific viewpoint. This started with Thomas Hobbes (Hobbes 1651) who made authority an absolute but founded on people's will, not as a Divine Right. He was followed by John Locke who favored Natural Rights, which included, "the inalienable right to life, liberty and property." (Locke, 1690) Locke was followed by many other philosophers who extended his concepts and formed what is now called "The Enlightenment" because it substituted logic and science for religious dogma and myths. These philosophies formed the basis for the American Declaration of Independence and the United States Constitution.

4.3 Other Philosophers

There have been many philosophers since who had different thoughts and which produced different ethics. This is no different than religious prophets who followed each other with new concepts and interpretations. Surprisingly, in practice, philosophy and religion have many similar characteristics because they share a common characteristic, of being a blind belief. They usually do not include the procedures of surveys, physical testing and confirmation by others that characterize science.

Philosophies however, differ strongly from religions, in that they change with time and circumstances. Also, older philosophies are often de-emphasized as the environment changes so as to make them inapplicable.

Except for the following items, this author has abandoned an attempt to analyze philosophical concepts/theories and compare them with SciEthics. It soon became apparent that would require another book. So, my emphasis on science should excuse this omission.

4.4 Natural Law/Common Sense

4.4.1 Natural Law

This author believes that the current philosophy of Natural Law is a continuation of the Enlightenment Era. There is a considerable literature under the label of Natural Law. The major items of Natural Law as defined by Edward J. Erler (Erler, 1984) and the corresponding SciEthics are listed below.

Natural Law/SciEthics

LIFE PL 1 Do what is necessary to stay alive.

FAMILY PL 3 Any progeny must be nurtured until they are able to be self-supporting.

THE GOLDEN RULE PL 4 Cooperate with and do no harm/hurt to others in your tribe.

LIFE LIBERTY AND THE PURSUIT OF HAPPINESS PL 5 Do your own thing but do not harm/hurt your neighbor. PL 9 You are free to go where the grass is greener.

REASON AND LOGIC PL 6 Value intelligence and creativity above strength and speed.

FREEDOM OF SPEECH PL 7 Speak up and give others the benefit of your thoughts.

THE EXCLUSIVE RIGHT TO PROPERTY PL 1 0 The things you make by your own efforts are yours.

No apparent basis for Natural Law could be found in the literature other than the philosophy of the Enlightenment and common sense.

4.4.2 Common Sense

Many people judge human behavior on the basis of what is called "common sense." Natural Law is a more formal academic exposition of common sense. The overlap between common sense, Natural Law and SciEthics is considerable. Common sense is the product of both nature and nurture. Sometimes it is correct in the SciEthical sense when it reflects genetically based instincts.

However, when it is based on nurture, the things it was taught by parents and teachers, it could very well be wrong in the SciEthical sense. The common sense that blacks were inferior and should be slaves was based on two thousand years of slavery and a self-serving apparent logic.

The inferiority of females, also based on 5,000 years of practice, was finally exposed as false by science, which substituted machines for strength and speed and thus allowed women to drive cars and trucks and airplanes. The advent of typewriters, telephones, electronic controls and computers also made women capable of operating in the commercial world. The advances of science and engineering freed women from the household tasks of antiquity and doubled the labor force available to commerce and industry without doubling the need for food, shelter, clothing and medical services.

This accounts, in part, to the wealth of the modern democratic nations that accepted and utilized the freedom of women.

Like philosophies, common sense is a product of the times/environment (as well as the genetic basis) and could be wrong in a deeper sense. It is often used to justify a position when there is no scientific basis. Thus, common sense should be accepted but used with caution.

PART B

Chapter 5 Science

5.1 Science in General: The Scientific Method

Science is a process of acquiring an understanding of how (not why) nature operates. In science, one uses observation, measurements, experiments and the construction of mathematical relationships and rules (hypotheses) that describe the observed behavior and predict new behaviors which are subject to confirmation by observation or experiment. Other independent scientists working in other laboratories must confirm these observations, measurements and experiments.

Publication of scientific findings or hypothetical consequences of the postulated rules must be truthful, open to all scientists and subject to confirmation or refutation. Reports must be in sufficient detail so that others can repeat the experiment. All terms and symbols must be defined. The instruments used must be so well described that others can replicate them. All the data obtained must be given so that the degree of accuracy of the compliance of the data with the theory can be judged. The suppression of data or the creation of false data is such a heinous crime in science it is sufficient to blackball the perpetrator from further participation in any field of science.

5.2 Non-Scientific Subjects

Not all subjects are susceptible to scientific study. If they are not capable of being examined, measured, duplicated by others, confirmed by experiment, etc., they fall outside the field of science. These exceptions fall into two main major categories; First, those that may eventually be possible but we have no tools to study them yet; i.e., emotions, mob behavior, intuition, etc. The second category is the impossible, i.e., absolute violations of scientific laws as two bodies occupying the same space at the same time, perpetual motion, the existence of God, miracles, etc.

5.3 The Amorality of Science

The results of scientific efforts represent a better understanding or a closer approximation to the understanding of nature. As such, they are amoral in themselves. They can be used for good or for bad purposes, ethical or unethical ends, moral or immoral purposes. Fire can be used to warm the home in winter or to burn "witches." Ropes can be used to harness horses for plowing or to hang blacks in the Old South. In any form of governance, the people in control make that decision.

If scientists are asked to be accountable for the use of their discoveries, then one is abrogating to them the responsibility and authority to make ethical/moral decisions which normally are made by the elected representatives in a democracy. Should the proud inventor/discoverer be allowed to decide whether his "baby" is good or bad? Also, there is no practical way to control what other people may do with the information. The chemist Nobel, who invented dynamite, thought it would be used in mining, construction, excavation of harbors and other beneficial purposes. He was horrified to see its main use becoming bombs and instruments of war. So, he used the profits from its manufacture to establish the Nobel Peace Prizes as his amends.

As an other example, with the discovery of nuclear fission, it was immediately recognized by the scientists involved that the energy released could be used as a bomb or as a source of heat to generate electricity. The decision to develop the bomb was made by President Franklin Roosevelt and the decision to use it was finally made by President Harry Truman; both elected representatives of the citizens of the United States.

5.4. Scientific Answers Are Approximations

Every scientific discovery must be considered as an approximation to the understanding of Nature and not an absolute truth akin to God's Commandments. Newton's Laws are not repealed or made false by Einstein's Law of Relativity. Newton's Laws still apply in our world. But for very small distances within the atom, action at velocities close to the speed of light, or for some cosmic phenomena, they require a correction.

In the same way when I went to engineering school we were taught two separate scientific laws; The Conservation of Mass and the Conservation of Energy. The discovery of nuclear energy in the late 30s demonstrated that mass could disappear and energy would appear. The two Conservation Laws were then combined into one, The Conservation of Mass and Energy. In 99.9% of engineering calculations today, the original two separate laws are still valid and are still used. Only in nuclear engineering does one make use of the combined law with Einstein's famous equation ($E = mc^2$) linking mass and energy. Thus, the original two conservation laws were not proven false; their region of applicability was defined more precisely. The approximation of the description of nature was improved. In this way, successive discoveries are mainly refinements or better approximations and should not constitute proof of the fallibility of science. While there are no absolutes, science provides the best description of Nature we have found to date.

Another example is the development of non-Euclidean geometry, which, while denying some of Euclid's basic premises, provides a comparable geometrical structure that can be recognized only by other mathematicians. The entire world still uses Euclidean geometry in all of its daily affairs.

The fault with criticisms that science cannot be trusted because its answers always changes is that it utilizes the attitude of the religious fundamentalists who believe in God's words as absolutes and not subject to change. This attitude is transferred to scientific findings. These critics transfer their modalities of religious beliefs in absolutes to science and insist that scientific findings also exhibit this unchangeable perfection.

Science however, with its methodology only provides approximate answers. As instruments and understandings improve, the answers become better approximations. Einstein's Theory of Relativity provides a better approximation to measurements at the atomic scale and when traveling near the speed of light. It also predicts the bending of light by gravity. The amount is incredibly small and astronomers neglect this effect except when light from a star passes very close to another heavenly body. All current satellite calculations neglect this effect. Today, 99.9% of engineering calculations are still done using Newton's equations. Another example is the Potolomaic Theory that the sun rotates about the Earth which was based on theology. Copernicus proved that the reverse takes place. This provided a more realistic picture of how the heavenly bodies move. But nobody reset his or her clocks as a result. To the great majority of the Earth's inhabitants, it really doesn't make much difference as long as the day is still 24 hours. Now however, we can correct the calendar every four years by inserting an extra day. Of course, it took 500 years for the Roman Catholic Church to admit Copernicus was right and they were wrong. This illustrates the difficulty of changing religious beliefs when science provides a new/better understanding of nature.

Still another example of the fact that scientific discoveries are approximations is the theory that the Earth was round. Many measurements seemed to confirm it. Yet, when we put satellites into a north-south orbit it turned out that the Earth is an oblate spheroid, slightly flatter at the poles than at the equator. Some one did a Newtonian type calculation on the centrifugal forces because of Earth's rotation and showed that because these forces are stronger at the equator than at the poles, the diameter at the equator should be larger than at the poles. Again, these differences are not enough to bother anyone other than a specialized surveyor or geographer.

Scientific theories explain reality. As a theory becomes more refined, it provides a better approximation both in concept and in numerical values. If one accepts the picture of science providing progressively better approximations, and then there is no basis for the attitude that each scientific advance proves the previous concept is wrong, therefore this one will be proven wrong in the future, and so science can't be trusted. This attitude reflects a belief in absolutes that is characteristic of religion.

5.5 The Evolution of Mankind

5.5.1 Basic Assumption

We accept as fact that the human race evolved by natural selection from a small tree-dwelling mammal. Modern man is designated by taxonomists as Homo sapiens. Other mammals currently alive also evolved from the same, or similar, predecessor. These are chimpanzees, apes, monkeys, gorillas, baboons, etc., Homo sapiens did not descend from today's monkeys. We diverged from a common ancestor long ago and thus we are all distant cousins. The degree of divergence can be estimated by comparing chromosomes and DNA chains. That's an interesting comparison but not germane to our brief history and will not be pursued here.

5.5.2 The Process

Let us follow the hominid as he left the forest and entered the flatlands or savanna. The departure was caused by the shrinkage of forests as the weather became drier. The food and water needed were on the savannas and in the lakes and rivers and no longer in the forests. Finding food and water was very important five million years ago, as it still is today.

On the savanna, he walked hunched over, knuckles on the ground but standing up frequently in order to look over the tall grass. He no longer had the ability to see great distances as from the top of trees. Over millions of years, evolution preferentially selected those specimens that could better stand upright to look for food and for enemies. This selection included those whose pelvis and hipbones rotated, whose stature straightened, and who were taller. The arms were no longer needed for balance on the ground, thus permitting the use of hands to carry weapons, tools and food long distances. This was a very highly selected configuration; no longer optimum for swinging from branch to branch but definitely more suited for savanna living.

Probably the greatest and most significant change was the increase in relative brain size as intelligence proved most effective in solving the problems facing the gradually changing humanoids. This intelligence led to tool fabrication and usage, the habitation of caves, the use of fire, and numerous inventions that reduced the mortality rate.

The social structure also changed. Initially almost all mammals had formed into small groups in which the males copulated with any female in heat (estrus). The dominant male usually, but not always, exercised his prerogative of first choice with a female in heat. After giving birth, the females supported and bonded with their offspring. The males established a dominance hierarchy and acted in concert to protect the group.

In the case of Homo sapiens, changes occurred. With the biological changes taking place, the period of an infant's total dependence on its mother grew and the ability of the female to feed and protect herself during pregnancy, and then the offspring over longer periods of time became difficult. Again, without any conscious design or goal, evolution selected those females whose sexual availability extended over longer periods than in the limited monthly estrous cycle. This was because the greater availability attracted the strongly sexed males who would bring her food, stay closer and try to keep the other males away.

From these factors, it is believed that families formed within the group. This may have been facilitated by the use of small caves as shelters that forced the larger groups to break up into smaller units during the night and during rainy seasons. While mothers could always recognize their specific offspring, now males could do the same. Again, mates who could recognize and preferentially protect and feed their offspring were selected. Thus, slowly, families were formed and the group of families became a tribe.

With the use of tools, clothing and weapons, food collection became easier and more efficient, providing even more time for non-basic activities. However, it still took up an individual's time and effort in order to fabricate these new devices. The practice of saving them after a usage must have grown, and thus the devices acquired a value. This led to the concept of private property.

The feeling that "I made it and it's mine" is very deeply imbedded in the instincts of man. The communist dream or slogan, "From each according to his ability and to each according to his needs" is very much against this human instinct. This conflict was a major factor in the failure of the communist experiment after only 70 years. (Other dictatorial and highly bureaucratic societies that accepted the practice of private property have lasted much longer, i.e., the Ottoman Empire.)

That slogan, to the extent it is followed, also in effect repeals the evolutionary process of selectivity because it broke the "cause and effect" of significant human characteristics. Of course the 70 years of communist rule was too short to have a genetic effect, but was long enough for basic human traits to reassert themselves.

As Nature (evolution) selected the adequately strong and aggressive, but more intelligent males, this began to establish what we now call masculine characteristics. In the same way, those female characteristics which attracted a supportive male, fostered the life of the offspring (mother love) and the effectiveness of its upbringing (care and training) were selected and became what we now designate as female characteristics.

There also occurred, in parallel with the development of families and male/female characteristics, a division of labor between the sexes. The men grew larger, stronger, more aggressive, hunted silently in groups that required a leader. The women grew more nurturing of children, gathering of fruits and vegetables, cooperative and talkative with the other women, doing those tasks compatible with the longer term care and protection of growing children still unable to fully take care of themselves.

In this context, adolescent females learned to attract males who would protect, feed and care for them and their children through the longer and longer periods of pregnancy and child bearing and rearing, now 8-10 years. Here also, she acquired those traits we now call feminine. If she did not acquire/learn these traits, the probability of having grandchildren would be less.

In this context, adolescent males were shifted from mother care to father care when they were physically able to follow grown men and imitate their activities; probably about the 10-12 age level. Again, if he did not acquire/learn these male traits the probability of having grandchildren would be less.

The first humanoids are now called Australopithecines and the earliest evidence of these tool-making, upright-walking mammals is about 5 million years ago. How long it took the Australopithecines to develop is uncertain. The subsequent appearance of new and more advanced hominoids are called Homo erectus and Homo robustus and finally Homo sapiens. The Homo erectus showed up about one million years ago. Thus, it took millions of years for the adaptations from tree living, branchiating mammals to upright, tool making and weapons carrying, more intelligent primitive man. This brief history is not a complete or academic discourse and is focused mainly on the genetic acquisition of "human traits."

5.5.3 Nature and Nurture

The distinction between nature and nurture, or genetic and cultural learning is admittedly fuzzy. Both contribute. From the behavior of small children, and of identical twins separated at birth, as well as the results of the communist experiment, it appears many more human traits or tendencies are genetically built in than were assumed by the proponents of culture as the predominant or even exclusive force in shaping human behavior.

Along with the biological changes came social changes. The forming of families with a specific pair of adults begetting specific children and protecting, feeding and training them for many years represented a major break with the life styles of most mammals and provided the evolutionary process an accelerated selectivity mechanism. Now in the main, a specific male selects a specific woman (or subtlety, vice versa). The offspring reflect the genes and cultural behavior of their parents. When combined, if these are superior to those of other adult pairs then there will be more grandchildren for them than for the others.

The social structure of tribes and larger groupings also goes through an evolutionary process of selection. Many forms of "government" are tried and those with inefficient features and which are "disliked" by their participants, self-destruct either by being abandoned, dying out or by being overcome and absorbed by groups or tribes with more effective forms of "government."

A group or tribes become stable and increases if their leader is intelligent and sees to it that under his direction, all members receive some of the benefits of the greater efficiency of the group in acquiring food, hides, furs, tools and some of the loot of warfare. This is an over simplified description. It does not cover the effects of the corruption of power, any bias in the selection of the next leader (i.e., his son even if an idiot), the need to split up the group if it gets too big for the resources of the local area, the conflicts if the tribe has too many potential capable leaders and the unforeseen results that often accompany a good intention. Our Intelligence today is still not enough to preclude or rapidly repair the unintended consequence resulting from the laws we pass.

5.5.4 Curiosity and Creativity

Some time during the growth of intelligence, as indicated by the growth in brain size, a primitive man some where, some how came up with the most brilliant invention of all time; one better than the spear or wheel or pottery or metal smelting or sliced bread. This invention was the framing of the question, "WHY?" or Y? and the creativity involved in trying to answer the question.

Mankind's curiosity, inquisitiveness, nosiness are summed up in the one word, "WHY?" In trying to answer the question, man created effective cooperative hunting strategies, learned how to swim, to not eat meat that had a certain odor, etc., etc. etc. Some times the answers were wrong but the negative feedback of failures prompted the formulation of a new answer or tactic or process. Failing a fundamental understanding, trial and error works quite well.

In order to answer the question, "Why is this here?" or "Why does this happen?" man invented gods with supernatural powers in order to explain that which, from observable data, he could not. He created many gods to explain why the sun and moon go across the sky, why people fall in love, why rivers run down hill, etc., etc., etc. To emphasize a point to be gone into in greater detail later, God did not make man, Man made God. Not until Darwin published his epochal, "Origin of the Species" did man have a scientific basis for knowing that "Why are we here?" is the wrong question. Attempts to answer it have been made for more than 5,000 years without great success. We now know that the right question is, "How are we here?" and the right answer is, "Evolution."

5.5.5 Science

In the attempt to answer the question "WHY?" many false paths were taken and many dead ends encountered. Finally a fruitful technique called the scientific method was developed. In this method a theory or answer to some question, "WHY?", is framed using observable physical data, repeatable by others in different places and subject to independent critical tests of the implications or predictions of the postulated theory. Mathematics is often the heart of this method. Mathematical representation of the theory permits extrapolation to other conditions and situations that can be tested to confirm or deny the theory.

The scientific method requires publication of the theory and the data in its support. There must be complete disclosure of instruments used, materials employed and processes utilized as well as all the data, so that others can independently repeat and confirm or reject the theory. Secrecy or the concealing of "proprietary data" is verboten. Any scientist who falsifies data, is caught lying or conceals details so that others cannot duplicate the experiment is "excommunicated" from the scientific community. Sure, some scientists do this, but the number are infinitesimal compared to lawyers or brokers or politicians.

We do not want to imply that no valid answers were found before the scientific method was developed and used. There were many very brilliant men throughout the ages who intuited the correct answers to many questions. For example, the question "How do we live together in large groups?" was answered in large part by Moses with the Ten Commandments and by Jesus with the Golden Rule. In addition, the method of trial and error is effective if one is willing to use the feedback of actual results to correct or modify an answer. Of course, if the answer had been supplied by a supernatural god, it was rather difficult to change it.

Thus, in the evolution of man, a relatively new tool, the scientific method, provides valid answers to the many physical "WHY?" questions in the world of nature. Lately, this same method is beginning to supply valid answers to "WHY?" with respect to mental questions.

5.5.6 DNA

A more recent event that provides information on human evolution is the discovery of DNA, the structure of the genes contained in the 46 chromosomes that are found in most cells of the human body. In the process of procreation, the male and female genes are each split in half in an arbitrary fashion. The female half is implanted in her ova and the male half is implanted in his sperm. Thus, upon sexual intercourse, the male sperm enters female ova and both halves combine to form a complete set of genes. Each child is thus a product of both parents. (The process is not always perfect and sometimes genetic deficiencies occur.)

The degree of evolutionary divergence can be estimated by comparing chromosomes and DNA chains. It has been estimated that 95% of the genes are identical between mankind and the chimpanzees. There are many interesting studies on the ability of the chimpanzees to learn sign language (they have no larynx and physically cannot talk), exhibit emotions and use tools.

By the ability to identify genes and to relate them to human characteristics, both physical and mental, it is now becoming potentially possible to alter the genetic makeup of an individual to alleviate and or cure certain ailments. This is designated as genetic engineering. Its first applications have been in agriculture, making plants more resistant to damage and more positive in product size, taste and ease of distribution with reduced spoilage.

Genetic engineering of humans is starting to develop. Cures for diseases are generally accepted, but genetic engineering to improve looks and abilities are very troublesome ethically. This topic will be discussed in greater detail later in Part D.

5.5.7 Deficiencies of Biological Evolution

Biological evolution works by selecting traits that improve the ability of the individual, and his/her descendants, to live better and to adapt to a changing environment. Homo sapiens' success is based on the behavioral traits defined herein as Primordial Laws 1-11. There was no conscious intent in their selection. The traits selected were in no sense perfect. They just worked better than previous traits. However, this selection process has several deficiencies.

The will to live as discussed more fully later is reflected in PL 1. This desire is so deeply imbedded in our psyche that it has no limits. The value of staying alive becomes less when the children become self-supporting. Then when the grandchildren arrive, grandparents serve as emergency baby-sitters and become an insurance policy should something happen to the parents. But when the grandchildren become self-supporting there is no basic need for the grandparents and their desire to stay alive represents an over kill of PL 1 as their function in the biological evolutionary sense is gone. There is value in what they can contribute to the societal evolution by virtue of the experiences and wisdom they have acquired. If they can no longer contribute to society, then even that value is gone.

Another example of over kill is the sex drive reflected in PL 2. After menopause, women cannot conceive and the sex act becomes recreational. Men do not have menopause or an equivalent and there is almost no end of their desire to have sex. Yet, there is almost no biological evolutionary purpose for sex after forty. The biological evolutionary selectivity process cannot operate if the individuals can no longer be procreative.

As the selectivity process is without plan or purpose, many physical aspects of Homo sapiens are very poor. Why does the appendix, which has no current value and represents a health hazard, remain? We no longer have tails, but the tailbones still exist. No civil engineer would plan to use the same channel for waste disposal and procreation. The evolution of the spine (backbone) from that which evolved for a branchiating mammal is poorly adapted to upright walking. Biological evolution is very slow.

Thus, the deficiencies of the biological evolutionary process can be summarized as, over kill, haphazard and slow. Does societal evolution act as a substitute selective mechanism for biological traits? See Note 5 for an example of societal selection.

5.6 The Evolution of Society and Technology

5.6.1 The Process

Probably the first characteristic selected by evolution by virtue of its effectiveness is the ability to act in concert with other individuals. There is strength in numbers and this was used by almost all mammals, including the ancestors of Homo sapiens. This group usually included several males and many females, young and adolescent children, and elders past the breeding stage. Usually one male was dominant, but that is part of the story of mammalian social development. This formation of groups or bands proved very effective in the preservation of almost all mammal species, and we see it today as a wide-spread practice, with many different forms from monogamous pairing to large herds and even to pods of whales for those mammals who went back into the sea. It is not the intent of this section to dwell upon the many forms of groups and how they grew into tribes and eventually into societies and nation, nor why and how this was evolutionarily beneficial to survival, which it obviously was.

The group of interest here is the family social group comprised of a male plus a female, their children, and grandparents. Siblings and their offspring would constitute another family. Multiple families would become a band and the bands would grow into tribes and eventually societies and nations. (This growth was evolutionarily beneficial to survival. How/why Homo sapiens formed families instead of groups is treated elsewhere.) Thus, we now have nations to consider.

5.6.2 Nations

If one follows the history of countries, nations, and societies, it appears that evolutionary forces are at work here too. Tribes grew larger and become city-states, then nations and now appear to becoming a global entity. This was accomplished more by war (Note 7) and conquest than by diffusion and assimilation. Many societies became extinct and others adapted to the environment established by their conquerors. Thus, political entities changed their form as well as their size. A thorough exposition of this view would require another and much larger book. Thus, only a few observations will be made here on that topic in order to illustrate the basic concept.

Societies have a tendency to become corrupt and ossify as the leaders solidify their control and increase their personal aggrandizement. "Power corrupts and absolute power corrupts absolutely." Soon, those in power consider themselves godlike; their pronouncements cannot be questioned or changed. Once everything is running to their own satisfaction, they have a perfect world and why change anything? Thus, progress stops and society ossifies, which makes it vulnerable to invasion and revolution.

It may be that wars are useful from a societal evolutionary viewpoint. They break up a frozen society and thus permit alternative forms to appear and evolve. In a way, when we in the USA change governments every four or eight years we accomplish the end results of a war without the military waste and slaughter.

Of course, this turnover has its disadvantages too. One is the difficulty for the USA to have long-range policies that are followed from one administration to the next. Some of these long-range problems are: exhausting natural resources, i.e., petroleum now and coal eventually, certain scarce minerals, pollution of the environment, i.e., smog and contaminated water supplies, population density, i.e., cities so large many inhabitants grow up with out any exposure to nature. Applications of science i.e. space exploration, negation of the biological evolutionary process, i.e., saving every Homo sapien for as long as possible. If we change administrations every four or eight years there is always the danger that sound programs are abandoned i.e. energy independence started by Richard Nixon, continued by Jimmy Carter and abandoned by Ronald Reagan. Putting a "Man On The Moon" in ten years is an exception, but it was not followed up. Apparently, if a program is started and heavily financed for as much of eight years, the momentum may carry it through.

It is the belief of this author that the lack of long-range policies is a lesser problem than the ossification of a society. After all, the Republican administrations that followed did not undo many of the policies of the Democratic administration of Franklin Roosevelt. It is apparent that the Democratic Party currently does not advocate abandoning all the Reagan-Bush policies they opposed when first instituted.

The theme seems to be, to this author, that societies that violated the basic SciEthics established by Nature failed, while those that accommodated them did better. The driving force to societal growth was the population increases due in large part to the increased efficiency of food production. This was accomplished by the invention of agriculture, the domestication of animals used for food and for labor. This permitted the support of armies and governments, the members of which did not need to hunt and gather food.

The failure of the communist experiment after only about 70 years in the Soviet Union and somewhat longer in other countries is in large part due to the violation of SciEthics, PL 1, 5, 6, 7, 8, 9 10 and 11. Why should one work harder to make more widgets per day than another worker who made fewer but was paid the same? If the property you use is not your own, will you take good care of it? If your creativity enriches others but not yourself, will you continue to create? If your genetic research results are denigrated because they do not agree with the government's political philosophy, do you continue doing productive research? How effective was the Soviet government's support of research on acquired characteristics?

Possibly, the greatest factor in the failure of societies is the abuse of power by its leaders. "Power corrupts and absolute power corrupts absolutely" is absolutely true. As described later in Chapter 11, leadership is an essential element in group/tribe effectiveness. There has to be someone to define a strategy, assign responsibilities and divide the fruits of the successful joint activity. But unfortunately, this position of leadership lends itself to taking the best portion of the kill. From this simple beginning has grown the self-serving practices of chiefs, kings and dictators. They also increased the privileges and booty of enough people to form a palace guard that would make sure their leader would not be replaced even if he were inefficient and ineffective. From the Roman Empire to European kingdoms to current Asian dictatorships, the abuse of power is a fundamental characteristic although it eventually leads to their downfall.

Particularly ineffective is the practice of passing on leadership power to one's own children (per PL 3) regardless of their abilities. Changing the leadership of a country every four years by popular election, although not perfect, is a fairly effective way to avoid this problem.

The growing popularity of Western economics and democracy in terms of free markets, elections of leaders, freedom of speech and press and individual entrepreneurialship are really reflections of many of the SciEthics guides expressed as PLs. The success of the Western World in defeating the Soviet Union showed the rest of the world what was the more effective culture and they are adapting it in bits and pieces and slowly.

5.7 The Evolution Of The Environment

During the millions of years in which man evolved from a tree living, branchiating mammal to a ground-dwelling, walking, hominid mammal with his hands and arms free to carry food or weapons, Nature supplied the environment with which these animals had to contend. More importantly, this environment constituted the forces that, in effect, blindly and without malice or intent, allowed those characteristics of the hominids that enhanced continued reproductive success to be preferentially selected.

This environment included hostile predators, often other hominids, the weather, too hot or too cold for comfort, storms, lightning, volcanoes, glaciers, earthquakes, presence or absence of food, water, light, caves for shelter and everything we today call Mother Nature.

Hominids, and later even Homo sapiens, could do little to change these environmental factors. The beavers could do more in providing protective shelters by building dams to flood low lands and building protective cavities enterable only through underwater channels. So, man had to deal, in general, with the environment established by nature. Sometimes he could utilize caves, which were also provided by Nature but only if no predator was using it first.

The development of agriculture and animal husbandry allowed for a tremendous growth in population density. With the development of religion and then science, the ability grew to control more and more of the adverse effects of the natural environment.

At the present time, most of us no longer live in an environment established by Nature but in an environment established by ourselves. As population density increases, society becomes the major environmental force and Homo sapiens today are no longer contending with the forces of Nature. We walk about without any fear of animal predators; if it gets too hot we open a window or turn on the air conditioning; if it gets too cold we light a fire or turn on a gas or electric heater. Often we simply adjust the thermostat and really do not know or care what changes the temperature. If it gets too dark, we turn on the lights.

Earthquakes are still a problem, but a reduced one. We can now, within limits, build earthquake proof homes, bridges and other structures. We have not eliminated the hazard, but we have minimized the resulting damage from moderate earthquakes. An 8.0 earthquake will still be a disaster, but in that event the rest of the society will hurry over and provide assistance, again reducing the resulting damage and hardships. Volcanoes of course are still beyond our capability to minimize their effects. There is, however, the growing ability to monitor active volcanoes and warn people to evacuate. As an exception, in Hawaii news of an impending eruption causes a rush of people, not away from the potential danger, but to the top of the volcano on the upwind side in order to get a good seat to watch the show.

Probably the most important thing we have accomplished is the scientific explanation of why, how and where earthquakes occur and that volcanoes occur from natural forces inside Earth, and not because a goddess was punishing non-believers who would not sacrifice virgin females to her.

While we still cannot control the weather, we can track and predict it so that people in the path of hurricanes and tornadoes can head for the hills or storm cellars. The minimum numbers of deaths in the U.S. due to hurricanes, as compared to the numbers in adjacent countries that do not have hurricane watches, shows how modern technology and the wealth to use it can reduce the risk of natural forces. So, while we still cannot control the weather we can minimize the harm it could cause.

So, if the natural forces no longer predominate in our environment what does? The problem of the scarcity of food due to natural causes such as drought, insects, limited land or limited productivity has been solved technically. The U.S. pays its farmers to limit food production and stores enough food after each harvest to make up for a subsequent poor harvest. France subsidizes its farmers enough for them to sell their products abroad at prices that constitute "unfair competition" according to the U.S. Department of Agriculture. Nevertheless, people in the Third World starve because their governments forcibly transfer farmers from their normal productive farms and homes to new areas for political purposes (Ethiopia), holds the price it pays for farm products down in order to favor the people in the big cities, and forbids the farmers from selling directly to anyone other than the government. There is also greater harm from climate changes—desertification (desert growth)—and from burgeoning population cutting trees for firewood and to create more farms (Brazil).

There are many reasons for starvation in the Third World, but not from the lack of technical ability to grow food. Again, it is the ineptness of the governments and societies that leads to over-population and food scarcity, not the lack of knowledge. Governments, religions and social customs are difficult to change rapidly and thus they tend to enforce rules that were effective centuries and eons ago when the environment was quite different. "Be fruitful and multiply and inhabit the Earth" was a good rule in 2000 BC, but it is counter productive now.

Homo sapiens today, living in concrete cities, struggles against an environment that is mainly societal and not Nature. Bad governmental policies such as economics (socialism does not work), and lack of population control (China is an exception. See Note 3) due to religious forces that are thousands of years out of date. There is crime, which is the act of certain Homo sapiens preying on other homo sapiens because it is easier (there is no natural source of food free for the taking.) then creating wealth in a constructive manner, which requires education, training and the forgoing of pleasures in order to accumulate saving for investment capital. We are no longer afraid of the predatory wild animals, we are now afraid of predatory Homo sapiens who will rob, slay, rape, mug others, or of Homo sapiens who drive while drunk. Humans preying on humans does have a long history and is probably imbedded in our genes by now.

Another modern risk is the development and utilization of complex powerful mechanisms. Operation of these complex systems has no genetic basis and must be learned. Mistakes by airport controllers, railroad traffic controllers, police and numerous poorly trained, and perhaps low IQ people, can have devastating results.

Today the environment is mostly man made even in the Third World. Thus, we have the opportunity to consciously modify and hopefully improve the environment. However, there is a danger that it will be irrevocably ruined. That is, made unsuitable for that specie. But there is also the hope that it will be preserved and immeasurably improved. Homo sapiens can do either. The knowledge and tools produced by science are amoral. They can be used for good or for evil, for the benefit of his fellow men and women or for their detriment.

Ethics or guides for human behavior have been conceived in the past in a setting or environment largely established by Nature. If these ethics, operating in that environment, resulted in social improvements which were accepted as fair, reasonable, or deemed "inherent," they became imbedded in the social structure and culture. It must be recognized that this acceptance was of the whole and really does not necessarily imply acceptance of each individual practice. For example, a greedy leader may have been accepted because he also was effective. His greediness was tolerated as the price for that effectiveness. Many religious people do not accept every tenet of their faith.

When the environment changes, as it often does, some of these specific ethics can become dead hands of the past and can prevent adaptability to the new environment. The blind process of evolution however, can result in the death of a specie if it does not adapt.

When the world was young, the people were too few to change, modify or pollute the environment no matter what they did. The advice to "Be fruitful, multiply and people the Earth" was conducive to greater security and productivity, and thus became codified in the Judao-Christian religions. Now that the population of the Earth is 5 billion and will soon be doubling, this behavior is clearly counter-productive. Yet, the dead hand of the past is largely still effective.

The carrying capacity of the Earth is a fuzzy number. It depends on the state of the technology, the social/political/economic systems that determine storage, transportation, and sharing practices. While Malthus was, and still is basically right, the population density keeps going up. Agriculture advances just delay, but do not solve, this fundamental problem. No exponential increase can go on forever.

Do Homo sapiens collectively have the ability to adapt to the new stressful social environment or will he become extinct like so many prior species? Luckily, a democratic society provides a negative feedback mechanism that should stabilize the situation. But, who knows for sure?

PART B

Chapter 6 Natural Law

6.1 Common Sense

Many people judge human behavior on the basis of what is called "common sense." This reflects the culture of a given country at a given time but also includes differences of opinion and changes with time. Common sense is used whenever there is a new circumstance and there are no laws covering the situation. Academically, common sense is characterized as "Natural Law."

6.2 Natural Law

It is believed by this author that a more formal academic exposition of common sense is "Natural Law." There is considerable literature under that label. The major items of Natural Law as defined by Edward J. Erler (Erler, 1984) and the corresponding SciEthics are listed below.

<u>Natural Law/SciEthics</u>

LIFE PL 1 Do what is necessary to stay alive.
FAMILY PL 3 Any progeny must be nurtured until they are self-supporting.
THE GOLDEN RULE PL 4 Cooperate with and do no harm to others in your tribe.
LIBERTY AND THE PURSUIT OF HAPPINESS PL 5 Do your own thing (But do not harm your neighbor)

REASON AND LOGIC PL 6 Value intelligence above strength and speed.
FREEDOM OF SPEECH PL 7 Speak up and give others the benefit of your thoughts.

THE EXCLUSIVE RIGHT TO PROPERTY PL 10 The things you make by your own efforts remain your own.

LIBERTY AND THE PURSUIT OF HAPPINESS PL 9 Go where the grass is greener.

The overlap between common sense, Natural Law and SciEthics is considerable. No apparent basis for Natural Law could be found in the literature other than the philosophy of Enlightenment and common sense.

6.3 Customs and Manners

There is a large body of customs and manners in any society. These are rules of behavior that are not laws. Essentially, these are the informal guides to what is a proper, accepted and expected mode of behavior for an individual in the given society. These guides, when followed, facilitate a multiplicity of people to live close together, cooperate and enjoy the benefits of that society. The enforcement of customs and manners is usually done in a nonbinding fashion; first on children by parents during their rearing, by cutting remarks from an observer of a violation, and sometimes, severely, by ostricism of the individual.

Customs and manners change slowly with time. For example, it was customary for a man to open a door for a woman, but if one did not, one could get a disdainful look but not be arrested. With the advent of womens' rights, or "feminism," this practice is disappearing. Essentially, customs and manners are the guides to a proper, accepted and expected mode of behavior for an individual in the given society. They are ethics in the sense that they do define a person's proper behavior.

PART B

Chapter 7 Application of Ethics

Ethics usually apply only to the behavior of people that affect other people. This behavior is often defined by philosophical treatises such as "The Social Contract" or "The Marriage Contract," religions, laws, customs and manners. In general, SciEthics for specific situations are defined in the section dealing with Primordial Laws, PLs, in Part C of this book. However, at this point in Part B it is necessary to cover the topics of "rights" and "laws."

7.1 Rights

At this point we must define "rights." The innate characteristics of humans as developed by evolution and reflected in our genes (nature) are abilities or desires. We may utilize these abilities as best we can based on the education and culture (Nuture) of the society in which we live. But in the environment of raw nature, these abilities do not guarantee life, liberty or happiness. We may die before puberty, starve, lose a limb in battle or an accident, become prostitutes or slaves or forced to unhappily follow a strange god. When we are numerous enough to form a society and a government, and if that government makes and enforces a law, then that law ensures us a right to do a specific thing or to expect a specific benefit, i.e., that others will not hurt or harm us.

Also, Nature does not ensure that the characteristics selected by evolution are to be effective for each individual in any and all situations. Nature's Laws as they pertain to inanimate objects are usually well obeyed. A released object will almost always fall down. (Unless there is a hurricane that moves it sideways.) Planets will follow an elliptical orbit around a star until it blows up. But Nature's Laws as applied to living things applies only in general to the entire population and not necessarily to specific individuals therein.

There are plants, insects, herbivores and mammals. Insects will eat certain of the plants, herbivores and mammals. Herbivores will eat plants. Mammalian predators will eat every life form, even other mammals.

Lightning, floods, earthquakes and volcanoes will destroy certain individual lives. Nature does not ensure or guarantee certain rights, laws or privileges to each and every specimen of a given living specie. It simply selects those individuals who are doing the best job in staying alive and perpetuating the specie.

7.1.1 Human Rights

The Right to Life, Liberty and the Pursuit of Happiness accrues to human individuals only from a government that passes the proper laws and enforces them. (An inalienable right to "Life, Liberty and Happiness" is a hollow promise to subjects of a despotic dictatorship.)

The terms Natural Law and Natural Rights as used in the literature of philosophy and politics carry connotations that do not relate to Nature. Nature has only selected certain characteristics of living things because they, in general, were more effective than other characteristics in perpetuating the specie. Those selected characteristics help but do not ensure or guarantee that any specific individual in the specie will lead a full life and leave many offspring.

7.1.2 Animal Rights

The rights of animals to humane treatment is a current ethical topic. On one extreme is the absolute, i.e., do not harm, destroy or eat any animal. They are God's creatures, just as we humans are. On the other extreme is the equally absolute, i.e., they are only animals. They have no souls so we can do anything we want to them. What is the SciEthical resolution to this problem?

During the long evolution of Homo sapiens, we first ate small animals and killed large ones in self-defense. We still do this today. In California in 1999 we raise chickens, ducks, turkeys, cattle and pigs for food (also eggs where applicable) and can kill mountain lions in self-defense. This is a direct continuation of a multimillion-year process that led to our current status, is still applicable and is therefore acceptable.

Being descended from hunters and gatherers, we can survive eating only meat (Eskimos) or only fruit and vegetables (vegetarians). However, in the past one had to spend a lot of time gathering if one was to subsist only on plant life. However, our biology is now also based on eating meat, which allows us to go longer between meals and to store fat for periods of no available food. Much of man's advancements are based on the greater efficiency awarded meat eaters. In the course of human development, three practices developed that bear on this problem.

The first practice is the domestication of animals such as dogs, cats and horses for efficiency in hunting, agriculture, house keeping and travel. In some societies, these are also eaten. Variability in societal practices provides more choices for evolutionary forces to work on in the selection of the fittest society. In times of famine all animals, even humans, are eaten by humans. We will have to wait for this evolutionary selectivity to decide which practice represents the best adaptation to the environment, which fortunately is now mostly man created.

The second practice is the domestication of animals such as chickens, goats and cattle for eggs, meat and milk. This is an efficient extension of hunting for food. With more modern facilities, fish are also being domesticated and they are included with animals in this discussion.

The third practice is agriculture, the domestication of plants. Although plants are living things and were as much created by God as animals and humans, the immorality of eating plants has not arisen as an ethical problem in religious dogma. Justification of this extension of non-meat-eating ethics is difficult, but not impossible. There are some people who think the world would be better off if we were not here.

However, do animals have any rights? Does there exist a contract between humans and animals? Are there two such contracts; one for domesticated animals and a different one for wild animals? The strongest case can be made for canines, then to felines and finally to horses. Although all of these are able to survive in the wilderness without humans, the domesticated animals have gained in the ability to survive and perpetuate its specie by trading its abilities to humans for food and protection provided to it by humans. Certainly, many species of dogs, such as poodles, could not survive. This picture is fuzzy because the humans have bred species of canines that would probably not have occurred in Nature.

If there is a mutual benefit, then there is a virtual contract. This contract establishes the ethics of human behavior to animals, or at least to specific animals. Can there be a contract between two species that can communicate only by physical gestures and not verbally? Recent research in teaching chimpanzees sign language indicates there can be communication between species. Also, there are many symbiotic relationships between plants, and between plants and insects, and between different species of fish. These symbiotic relationships take on the character of virtual contracts.

When we look at the evolution of humans, a major step was the agricultural revolution in which man planted, fertilized, watered and gathered the cereals, fruits and vegetables for his food in addition to his usual gathering and hunting, which gradually diminished. I do not think there is a mutual contract between plants and mankind. SciEthics says that what man has done to bring us to our current biological status is ethical. This explicitly approves our freedom to harvest and eat plants. Vegetarians who avoid meat on ethical grounds avoid the fact that plants are also living things.

There are a large number of animals that are grown in a controlled environment for purposes other than food. Some expamples are mink for fur, rats for laboratory testing, birds for song and all animals to exhibit in zoos for the education of mankind. In man's long evolutionary growth he has utilized whatever he found in nature that would make his life easier and thus help improve and perpetuate his specie. Thus, the basic axiom of SciEthics, which says that what man has done to bring us to our current biological status is ethical, implicitly approves our use of animals for other purposes than food.

Thus, it is believed that a virtual contract does exist between humans and a limited number of animals which defines ethical human behavior to them. There is no such contract governing human behavior with regard to animals not included or to plants, insects and other life forms. There may be restrictions on human activity with regard to all life forms from ecological considerations of avoiding the elimination of species that might be found in the future to have a value to Homo sapiens. This aspect is treated in another section of this book.

The most obvious animal contract is between a man and his dog, which was domesticated about 140,000 years ago. (Allman 1998). Man is to a limited extent dependent on his dog for protection, as an aid in hunting and in traveling, as in pulling his sled. The dog in turn is dependent on man for his food, protection from the cold and for protection from larger predatory animals. In addition, the dog seems to be in need of affection from his owner. Since mans owns the dog, the animal contract treats the dog as a slave but no one seems bothered by that. The major concern by animal lovers is the treatment or neglect of the dog by its owner. On the basis that there is an implicit contract between the two, this concern is warranted and the dog does have the right to life and the means thereof, such as food and medical attention, but not to liberty or the independent pursuit of happiness other than sex and care of offspring. Since pain can act as a prelude to death, torture is also forbidden.

This right is extended to cats and horses in our society. No other animal appears to be included in any significant manner or number. Thus, the SciEthic resolution in the animal rights argument is that it's all right to raise animals and plants for eating purposes, but that animal rights are limited in our society to dogs, cats and horses, and are limited to their life and the means thereof as exemplified by PL 1, 2 and 3. This should not apply to other animals that are treated as household pets. The raising of a tiger cub in a household with children is an exercise in poor judgment. The tiger is not a domesticated animals. Neither is a python.

7.1.3 Corporate Rights

A corporation is a legal entity, pays taxes, obeys and is protected by laws. Is it a "person?" What ethical principles, applicable to individuals, are applicable to corporations? Corporations, if they are successful, can be immortal. This embodies PL 1 to the ultimate, which no person can do no matter how successful he/she is. But this immortality violates the evolutionary principle that members of any specie must die in order to permit continued selectivity. Corporations can also legally change their own basic function and character by intent, which is done in humans only by blind mutations. Should corporate charters be made more specific and unchangeable and have an end date? Doing so would make the corporate entity more like a person.

It appears that almost all of the PLs apply to corporations. By stretching PL 2 to include merging (marriage?) and illegal collusion between corporations (sex outside of marriage?), stretching PL 3 to include divisions and subsidiaries, PL 4 to include membership in legal trade associations, PL 5 to mean making a profit, then all the PLs appear to apply. But if SciEthics are to apply to corporations, they are responsible to do the same to each other and to individuals.

The detailed application of SciEthics to corporations will not be treated in this book. Anyone else is free to elaborate on this topic.

7.2 Laws

7.2.1 Evolution of Laws

The application of ethics thus far has been voluntary. Enforcement of ethics in primitive societies is usually by verbal criticism, orders by the chief or ostracization by one's neighbors. As society grew larger and more complex, ethics were then transformed into laws established by the rulers and reduced to writing so that all would know the same rules. Laws now represent ethics that are established by the government and enforced. This is an aspect of governance and is treated in greater detail in Chapter C 11 as part of PL 4, Membership in the Tribe.

7.2.2 Enforcing the Laws

In order to enforce its laws and induce others not to break them, society through its government exercises the power of punishment. Without punishment, laws would become meaningless. Actions against the well being of a given society as a whole or that of individual members would grow until the given society would disintegrate and its members disperse. Thus, the ability to punish and inhibit undesirable activity is considered a necessary characteristic of a government.

The punishment in today's society is usually directed at the criminal and in the form of loss of wealth, i.e., fines or confiscation of assets; loss of liberty, i.e., confinement in prison, or restricted activity while on parole; or rarely loss of life.

Although the laws in a SciEthical society would be somewhat different than in a religious or secular society, a crime would still be the same, i.e., breaking a law. The nature of the punishment would be somewhat different. Although still directed at the lawbreaker, the emphasis would be on providing restitution to the victim. In a sense, the lawbreaker and the society have not lived up to their contracts. The lawbreaker did not obey the law and society did not protect the victim. His suffering should be alleviated. This is not a morally derived conclusion. It is derived from the fact that if society does not uphold its end of the bargain, members will leave and go elsewhere or form a new society.

Thus, the thrust of punishment should be to undo the damage to the victim. This subject is developed in greater detail in Chapter C 11, PL 4.

7.3 Conflicts, Resolutions and Changes

With the many different religions and philosophies we are blessed with, there are often conflicts between the ethics or morals propounded. As explained earlier, there is no universally accepted Ethical Constitution that can be interpreted by a court system that you can turn to in a dispute. Neither is there a way, universally accepted, to repeal, modify or amend an ethic as of a given date. Changes in ethics do occur, but slowly and painfully over long periods of time.

For the last 2,000 years women were mothers, housekeepers and caretakers of the sick and elderly, but they were not warriors. (Were the Amazons an exception?) With the growth of science and engineering, skill replaced strength and women became involved in the military: driving vehicles and airplanes. (They had been involved earlier as nurses.) While this capability was used only in delivering vehicles in WWII, they became combat pilots in the Gulf War. Today, women are an essential part of all the military services in the USA. Thus, ethics change. Morals change more slowly, but they also change. An example is the attitude toward sex in the 1960s. The driving forces in these changes are science and technology.

But when ethics and morals are interpreted by laws and enforced by the government, only voting on propositions or public demonstrations can influence the government in a democratic country to change the laws. In a non-democratic country, the people just suffer or vote with their feet by illegally emigrating, or revolt.

PART C ETHICS BASED ON SCIENCE-SCIETHICS

Chapter 8 PL 1—The Will To Live

8.1 Life

The desire to stay alive is evident in the struggles of a drowning person, the frantic speed of a prey fleeing a predator, and the twisted tree growing out of a small crack in a cliff. The determined efforts of survivors of a sunken ship to stay alive on a raft waiting for rescue while enduring incredible hardships is the classical example. Snowbound stranded travelers have resorted to cannibalism. The desire of prisoners and slaves to stay alive although their social environment may be very stressful is another sample of the will to live, a genetic trait. Suicidal actions are an example that there are no absolutes.

The immense strength of this genetic trait, the will to live, is apparent in all species. However, Homo sapiens have extended this to the concept of immortality. This concept was a natural outgrowth of two traits selected by evolution: the will to live (PL 1) and intelligence (PL 6) which provided the ability to look ahead or imagine the future.

We have habitually planned ahead; as in hunting practices, aggression tactics and defense precautions. When we get older, and as death approaches, we do not want to die, but realize from the experience of others that die we must. Thus, we invented the concept of life after death or immortality. Evidence of this is strong as food, weapons, clothing and even servants are found in Egyptian burial sites.

The entire concept of burial rites, found in almost all primitive societies, implies that the concept of life after death is widespread. The knowledge that dead bodies have no future, get eaten, rot or disappear must have been gained very early in evolution. The attempt to preserve the human body by burial, whereas no other mammal does this, indicates an anticipated usage of the body in another lifetime. This desire for continued living was utilized by the founders of religions whom promised living forever in a heaven provided by their GOD concept, but if and only if you believed in HIM and followed HIS rules. This concept of immortality is widespread even today, illustrating the strength of the will to live, PL 1.

8.2 Death

8.2.1 Definition of Death

Historically there have been three definitions of death. When tribes wandered about, hunting and gathering, an individual who broke both legs for example, or became blind, was functionally dead. This individual was left behind to die as the tribe moved on. Thus, an earlier definition of death was, "Any disability (which included the infirmities of old age) that prevented the individual from traveling with the group." By virtue of habit, custom, and religious practice that definition is still used today by many primitive people.

With the advanced environment of agriculture and cities or fixed abodes after the hunter/gatherer economy, death was long defined as the cessation of heart and lung activity when there was no recovery possible from that condition. So the patient was buried, everyone said the proper prayers and some went into mourning. During the last 15,000 years, that definition was the proper one.

In the last hundred years, science and technology changed that definition. Machinery that could keep the patient alive after his/her heart and lungs had stopped and intravenous feeding were developed. Thus, the patient that would have died, by the old definition, several decades ago can now be kept alive indefinitely by that old definition. But once the environment changes, then our definitions must be reexamined and, if necessary changed.

Thus, a new definition was formulated based on brain activity. If a patient is 100% brain dead, without any chance of recovery, then he cannot behave as a human, is functionally dead and is so defined. If a patient is not 100% brain dead in that the primitive brain stem is still working but the cognitive brain is dead, without any chance of recovery, he/she is functionally dead and therefore this condition is included in the new definition of death.

This definition would include the condition of newly born encephalitic infants who do not have a cognitive brain but can be kept alive by machinery. Under the new definition of death, they should be allowed to expire by whatever means is acceptable to the mother or family. This could be by injection, smothering or absence of nutrients. Garnishing organs from breathing infants who are dead by definition is ethical if it will keep a living person with a fatal illness alive.

The clause, "without any chance of recovery" is not meant to be an escape clause. Some patients will be kept alive in a brain dead condition by family choice until they recover or die from other causes. Data can/should be analyzed to determine the longest time beyond which there is no hope of recovery. This may have to be stated in probabilistic terms that may be difficult to understand, but will establish a time when the plug should be pulled on a patient that is being kept alive by machinery but is dead by our new definition of brain dead.

If one insists on using the old definition, then a simple alternative procedure would be to place the patient in the environment that existed when the original definition of death was established. (No respirator, no intravenous feeding, no drugs not available in Biblical times, etc.) If the patient lives, he is alive, if he expires he is dead.

From a scientific viewpoint, there is no need to struggle with the religious sophistry of the "sanctity of life" or "a person's soul" or "only God can take what He has given." If a person has no probability of ever being able to act as a conscious person then that person is dead.

8.2.2 Why Death Is Necessary

The great strength of the will to live is written into our genes. Although the incidents of suicide are exceptions, these are excused by the absence of absolutes in science. Thus, the will to live applies to the great majority of people.

This will to live has given rise to many attempts to create eternal life. This was first assigned to the gods, but later extended to humans by the creation of the concept of heaven where life was eternal. One might have to live a certain way to get there, as defined by the Wise Men of the tribe, later by clergymen who established different rules in the name of different gods. For the Norsemen, to die in battle was to go directly to heaven. Then there was the search for the Fountain of Youth in the New World. Even today, the concept of immortality is accepted by Islamic terrorists who self-destruct in bomb attacks.

However, the science of evolution implicitly requires that all specimens of any specie die so that they can be replaced by specimens better adapted to the environment. If the first organism that could replicate itself had been immortal, the ocean would be full of a simple (single cell?) life form and there would be no plants, fish, animals or humans. Because the single-cell life form procreates by division, part of the first single-cell specimen might still be in the ocean if it had been adaptable to all the environmental changes during the last billion years. (Not highly probable.) Death of our ancestors was necessary for us to get here as shown in Figure 1.

Without death there would not be any life chain where one specie lives by eating a lower form. No herbivores eating plants, no predators eating herbivores, and no humans eating anything available.

Only life forms that absorb inorganic molecules would be present.

Even if the concept of immortality were limited to Homo sapiens, how would fatal accidents, murders and terrorist bombs be avoided? Of course, if immortality is limited to those in Heaven then there could not be any of these incidents. Death is necessary and unavoidable, period. The long-lived concept of immortality is due to two factors:

a.   The basic will to live embodied in our genes.

b.   The intelligence, imagination and creativity which generates apparently possible desired solutions to life's problems.

8.3 Applications of PL 1

8.3.1 Murder

Murder is an absolute violation of PL 1 and is an unethical act. Attempts to justify it on the basis that it accomplishes a greater good is unacceptable as there are always other means to achieve the greater good and selecting that greater good can be very subjective. This argument would be used to justify political assassinations, terrorist bombing and even genocide.

The Sixth Commandment (Figure 3) is significantly different in the Christian and Hebrew versions. The Christian version says, "Thou shall not kill," while the Hebrew version says, "Thou shall not murder." Today "kill" means to cause the death of a person or animal while "murder" means unlawfully killing a person with malice aforethought. Thus, any warfare creating fatalities is prohibited in one version but not in the other. However, what was the meaning of the two words at the time The Bible was written and/or translated?

8.3.2 Suicide

There are many reasons for committing suicide and there is no simple answer as to its ethical content. It is not always unethical or always ethical. From the viewpoint of the continuation of ones genes, the suicide of a teen-ager or young adult is a terrible loss to parents and to society. From the viewpoint of society's investment in the birth, training and education of an individual, suicide is unethical. But PL 5 implies that the individual is free to do as he pleases as long as it doesn't harm others. As a person gets older, has had children, has repaid society and then gets painfully ill, his suicide if still competent is not unethical. Each case should be judged on the circumstances. There is no absolute yes or no.

8.3.3 Self-Sacrifice

There are many circumstances where people have sacrificed their own lives in order to benefit others. The practice of preferentially saving women and children when a ship sinks reflects the willingness of the men to be sacrificed in order to do so. This decision is a reflection of the primordial societal drive for the preservation of the family and tribe. The more women and children that are saved (and a few men to row the lifeboat) the greater the probability that the tribe will continue. (See Section 8.3.7.)

With the growth of society and the identification of oneself with a larger group, then the sacrifice of oneself for the success of the nation has become a useful cultural trait. Examples of such heroes range from the handful of Greek troops defending the pass of Thermoppoly to Ethan Adam Hale regretting that he had only one life to give for his country. As long as the people saved have enough similar genes to the "hero" then he has contributed to the continuation of his genetic composition and his self-sacrifice serves a useful purpose from a biological viewpoint. Even if there is no biological basis, societal basis exists and justifies his action. Yes, self-sacrifice can be SciEthical from both Nature and Nurture viewpoints. But the benefits must be real and not imaginary.

8.3.4 Euthanasia

If the person has requested it, is competent and is suffering from a painful incurable ailment, then it is SciEthical and in compliance of PL 5. Dr. Kevorkian's actions are perfectly SciEthical. He has been very careful to observe the above conditions. Oregon has passed a law to allow physicians to assist in such endeavors but with precautionary steps. But killing another person on the basis that it is beneficial to the community is unSciEthical.

8.3.5 Abortion

The current major argument against abortion is religious, that the fetus has a soul given by God and is a potential person from the moment of conception. The fact that approximately 50% of all conceptions self-abort in the early days of pregnancy is explained away by "God moves in mysterious ways his wonders to perform." If one is deeply religious, then there is no way to refute this position. The current religious position is that a person is inherent in the fetus as soon as the ovum is fertilized, and that abortion at any time is the murder of that person.

The counter argument for a woman's right to abortion claims that she has the sole right and responsibility of deciding what to do with her own fetus/body. Let's assume that this abortion right is limited to the first trimester and that this time represents a safe dividing line between non-personhood and personhood. This is the Wade vs. Roe decision of the Supreme Court and is based on the fact (at that time) that a fetus could not live outside its mother's womb until the second trimester. On this basis, the only people involved are the woman and her doctor. The reason for the abortion is not involved.

In the Jewish tradition, a person is born only when he/she leaves the mother and is living independently. Thus, the anti-abortion arguments depend on the definition of when does a person start. These arguments, pro and con, are cultural and not based on any scientific fact. Let us examine how SciEthics would treat the subject.

Many abortions occur by themselves in the first three months of pregnancy. No great fuss is made about such incidents, which are often unreported. These unplanned abortions are Nature's way to eliminate defective fetuses.

However, societal considerations also establish conditions under which some fetuses are undesirable for non-biological reasons. These include unmarried teenagers, too many children, interference with the completion of education, unavailable father and foreseen economic inability to raise the resultant baby. While completion of the pregnancy can be advocated on the basis that the infant could be given up for adoption, this course has many disadvantages, economic and emotional.

In many of these situations, continuation of the pregnancy would result in violation of PL 3. In view of the current high population density, the loss of a potential person is no longer critical to the continuation of the tribe. Although an apparent violation of PL 1, deliberate abortions, if desired by the female are SciEthical.

In rare cases, the father's desire and ability to assume responsibility to raise the child becomes a factor. Deliberate abortions are society's (Nurture) imitation of natural (Nature) abortions. The greatest prevention/restriction to abortion is the woman's innate genetic desire to have children and the mother love for the resultant infant.

8.3.6 Infanticide

Where abortions are not feasible, the killing of a newborn is often utilized. While many of the reasons for approving abortions are no longer applicable, some of the considerations detailed above for abortions still apply, but PL 1 becomes predominant.

Whether a "person" is existent at the moment of ova fertilization (Catholic), when the fetus can survive outside the womb (Supreme Court), or when it leaves the mother's womb at birth (Hebrew) is a legal/religious viewpoint. From an evolutionary viewpoint, if it is a random event, it is a continuation of the abortion process. If the infanticide is non-random, i.e., mainly of female babies (China), that disturbs the male/female ratio of the population, it is unSciEthical.

The question will be asked, at what point of growth, at what age, does infanticide become murder? It would seem that if infanticide is an equivalent of abortion, that it must occur at the mother's expressed wish and within an hour after the birth, well before there is any bonding.

Another question that will be asked is what if the infant is born defective in some serious way. With modern technology, this can usually be determined well before birth and an abortion performed. But if it is determined only after birth, then again with the mother's informed consent, it should be considered a terminal abortion.

8.3.7 Women and Children First

An interesting ethical problem arises when a ship is slowly sinking and there are not enough lifeboats for all the people aboard. This problem was vividly portrayed in the movie Titanic. Who gets to be in the lifeboats?

The nautical rule for centuries has been "Women and Children First" but after enough crewmen to handle the boat. If there is no crew, the lifeboat is probably doomed and no one will/may be saved. This is very dependent on the weather. The Titanic situation was rather unique in that the sea was calm, but even then, there had to be crewmen to row and steer the lifeboat away from the sinking Titanic or they would have been swamped when the ship finally went down. Usually a passenger ship goes down in a storm and there has to be a male crew in the lifeboats.

The preferential saving of women and children reflects the implicit goal of evolution to preserve the specie. A multiplicity of women is optimal for bringing new members into the tribe. Men are essential too, but the best route for survival is to mostly save the women and children and then a few men. In a very severe situation, i.e., with inadequate food and warmth, sacrificing the children is preferable to sacrificing the women as they can produce replacement children. Thus, evolutionary forces have in essence created a proper ethic, "Women and Children First."

Like all evolutionary ethics, this one was formed over a million years of small primitive tribes living in a world empty of Homo sapiens and in an environment created by Nature. This environment has changed drastically in the last 10,000 years since the development of agriculture and the changes have accelerated in the last 300 years because of science and technology. These changes are allowing women to become, although slowly, equal to men in the political, social and economic fields. A field of athletic activities for women has also bloomed, so it is probable that in a group of women in a lifeboat these days, some could row and steer it as needed. And if feminism is today's culture, then why not have men and women, in equal numbers, in the lifeboats after the children who would still come first?

Science and technology have increased the production of food so we can grow more food than is needed. The shortages of food in Third World countries is mostly due to their political and economic systems which reject the western systems that have proven more effective in most of the world. Science and medical technology have improved fertility of women, prevention of some diseases, effective treatment of others and increasing women's fertility period and total longevity of both men and women. The sum of these advances is an increase in world population unimaginable 300 years ago.

Thus, the need to save a few women and children in order to ensure continuation of the specie is not applicable in today's environment. But if not women and children first, on what basis does the captain of a sinking ship allocate room in a limited number of lifeboats?

As has been described elsewhere in this book, the major factor in the advances of humanity has been the selection of intelligence, PL 6, and not male strength or female fecundity. Thus, the basis of selection should be the intelligence of the passengers particularly, but not exclusively, those still of child-bearing age.

As an extreme example to illustrate this point, assume that Albert Einstein, Charles Darwin, Isaac Newton, etc. were passengers on a sinking passenger vessel with hundreds of other nondescript men, women and children. For humanity's sake, should not these highly intelligent scientists be saved first? Some may argue that the practicality of measuring the intelligence of passengers while the ship is sinking makes these criteria immaterial at the time. However, passports could be required to include the IQ of the bearer. The captain could, with the help of a computer on takeoff, generate a priority list of who is to be saved in the unlikely but possible event of his ship sinking. All this of course would require many changes in our culture and society, but the value of intelligence is much greater than the sheer numbers of new persons who could result from children growing up and from women bearing more children. That criterion is no longer applicable in today's society, but it is highly unlikely that this application of SciEthics will be adapted. The major purpose of including this idea is to encourage the reader to examine all ethical problems from the viewpoint of SciEthics.

8.3.8 Cryonics and Immortality

The concept of immortality reflects the strong evolutionary selected desire to live and is another example of overkill. However, humans have recently created a plausible method of obtaining "immortality." This is to freeze the body shortly before or after death (cryonics) and preserve it until future medical science has advanced to the point that the body can be resurrected and cured of whatever caused the death. Then one can see what the future is like and after a while can be frozen again, this time for a stated period of time. In this manner one can live a year or two every five hundred or a thousand years and see what the future holds for mankind. With the advent of electronic records of what happened in the past, there is no value to this periodic resurrection and it serves no evolutionary benefit. It is not SciEthical other than an exercise of PL 5.

PART C

Chapter 9 PL 2—The Sex Urge

9.1 The Basis of Sex

Bisexuality, the existence of male and female who have to provide semen and ova in order to propagate the species, is a complicated system but it provided many variations in genetic composition and characteristics so that selectivity had more options. The division of microbes by splitting in half is much simpler but does not provide the variability resulting from bisexuality. This is another example of how complicated systems develop by evolutionary selectivity. We will now focus on the functions of sex in our specie, Homo sapiens.

The strong sex urge in both man and woman acts to ensure that they will copulate and produce offspring. Thus, the desire for sex is very fundamental and sound. Without it, you and I might not be here. Thus, sex is good and not bad. Sex also plays a strong role in helping compliance with PL 3, the nurturing of progeny until they are able to take care of themselves. Among most mammals and our simian relatives, copulation only occurs when the female is in estrus, which occurs for only several days once a month. The male can sense this by the scent given off, changes of color of the genital areas of the female and her receptivity to his attempts to mount her. Thus, the male has sex about once a month with a given female until she becomes pregnant. Then there is no sex with that female until her first estrus after the birth of the offspring. This may not take place for some time while she is suckling the young one(s). This is not much sex for an adult male so he looks for other females who may be responsive. Thus, the male may have sex many times during the month or even in one day, but generally with different females. There is no bonding between the male and any one female. There is a polygamous bonding with all the females in the group.

As far as I know, Homo sapiens is the only specie where the female practices sex for the entire month between menstrual periods, which has replaced the estrus periods, and during pregnancy. It is surmised that this change was selected by evolution because it kept the male attached to the specific female by the frequent availability of sex over a much longer fraction of the time. There is no biological purpose for sex during a female's pregnancy. Its only function is to keep the male around and protect and feed the pregnant female as her mobility decreases. While the mother is still breast feeding and taking great care of the young, which in the human case takes many years, she needs help and protection. Thus, sex is the basis for the practice of marriage. The quick resumption of sex after birth again helps keep the male attentive. Thus, sex serves other useful functions than simply fertilizing the female ova during an enjoyable exercise.

The basic quid pro quo has been distorted in many societies. The laws of the Catholic Church that forbid sex except for the purpose of procreation are unenforceable. The great bulk of the Catholics in the Western World practice birth control, which is against canonical law. In countries where birth control technologies are not available, the women have too many children to care for them properly. In addition, she is too burdened to provide the male the attention he needs. This may lead to maltreatment of the spouse and often of the children who are seen as the cause of the lack of attention.

The temporal over kill of the sex urge is another example of the blindness or lack of intent or design in the force of evolutionary selectivity. By over kill is meant that the urge to copulate exists from approximately age 13 to approximately 75 (62 years), which is much more then what is needed to perpetuate the specie. Sexual activity between the ages of approximately 16 to approximately 40 (24 years) would be adequate. The duration of sexual urges is almost three times that needed for specie propagation. There is a secondary function of sexual intercourse in that the binding, via love, of the marriage contract in that the pleasure of the intercourse helps make the man put up with all of the female's idiosyncrasies and helps the woman put up with all the male's forgetfulness.

The major point is that to ensure the continuation of specie, evolution favored those individuals who exhibited a strong sex drive, which became imbedded in the genes. Suppression of this strong sex drive may lead to psychological harm to the individual, yet such suppression is widespread. The reasons for this are developed in the next section.

9.2 Sexual Restrictions

With the strong urge for sexual intercourse described above, why are there so many social and religious restrictions as to its exercise? Why is it deemed illegal, immoral and socially forbidden except within the context of the marriage of a man and a woman? There are some primitive societies where early sexual intercourse is not forbidden but allowed so that a female can demonstrate that she is fertile and can bear children. This makes her acceptable to a potential mate. (Benedict, 1934) But in most societies, sex before and outside of marriage is forbidden. Why?

First, this was not a serious restriction when for many centuries marriage was started at younger ages than today. Marriage soon after puberty, at 13 to 16 years of age was common. Extended families lived on farms and children of that age could be economically productive. In addition, the extended family provided care for pregnant girls and more senior judgment and direction for bringing up the resultant babies.

The problem with these restrictions has become more serious as the amount of education has increased and young people are in high school until 18, college until 22 and professional schools until 28. In addition, the population on farms has shrunk from 80% to about 5% in the U.S. Child labor in factories off the farm is completely inconsistent with the evolutionary history of child upbringing. Thus, the use of child labor in industry starting with the industrial revolution has generally been abolished in 150 years.

There are several strong reasons for forbidding sexual intercourse in the early puberty years based on the evolutionary process as defined in Sci Ethics.

These reasons are as follows:

Unwanted Pregnancy. The consequence of casual sexual intercourse is often a pregnancy. However, the very young person is usually unable to take proper care of the baby and there is no father to serve his basic responsibility to protect and feed the child and its mother. Thus, the responsibility as defined in PL 3 does not exist and the future of such babies is bleak. It represents a lower probability of successful propagation of the specie.

Virginity. The instinct for mother love and father love is based on the baby sharing the genes of its parents. While there is no doubt as to who is the mother of a newborn, the father could always be in doubt. This is reflected in the pressure for a female to be a virgin until marriage. In this way, the husband was assured the child was his.

Diseases. Casual sexual intercourse outside of marriage is more subject to sexually transmitted diseases. Thus, the success of genetic transmission to subsequent generations is enhanced by sexual fidelity within a marriage. While pre-marriage sexual activity by the male is generally accepted in most societies, this represents the strong sex drive of the male permitted expression in a patriarchal dominated society. From basic SciEthics, the propagation of the genetic makeup would be enhanced if both the male and the female were virgins.

Adultery. The marriage bond and the family bond, depending on the father being assured the child was his, forbid adultery on the part of the woman. Adultery by the male could result in a child he could not support. The danger of sexually transmitted diseases from adultery by the male is also a potential hazard. Thus, the evolutionary process is potentially violated by the adultery of the man or the woman.

All of the above persisted for millions of years during the evolution of Homo sapiens. However, with the development of 20th century science, the situation has changed. Now by the use of DNA signatures, a male can confirm that a child is his or not. The solution to the problem arising if it is not his is difficult. DNA testing to confirm the father is not yet widely done but is available.

Couples can be tested for sexually transmitted diseases before marriage. The use of condoms and other practices are called "safe sex" because they generally prevent disease transmission. There are many post-coitus medicines that will prevent accidental pregnancy. Thus, the environment of sexual intercourse has changed rapidly. Science now permits sexual intercourse without most of the undesired evolutionary consequences. However, the psychological consequences are still deeply embedded in the psyche and violations of the ancient codes of sexual behavior still have consequences. This aspect is discussed in greater detail in Part D.

The intent of this portion of the chapter on sex is to show that the religious codes are not merely arbitrary but do have a basis in evolutionary SciEthics.

The strong sex drive in male mammals is reflected in the social organization of a collection of individuals who lived together and cooperated in hunting. There was one strong and dominant male and many females with their offspring. There were some younger and non-dominant males. This social organization is quite effective and we see it today in many species. This has been carried over to Homo sapiens in the practice of polygamy, which is discussed in greater detail in Chapter 10 on PL 3. The majority of mankind today however, practices monogamy, which is more effective according to SCI Ethics.

9.3 Aggressiveness

In many of the mammalian species that pre-dated and in many that parallel Homo sapiens, conflicts between males is prevalent as a method of evolutionary selection of the dominant male who then impregnates the willing female during her estrus period. This is a selective process because genes of the most powerful and aggressive male are passed on. Thus, the males become bigger, stronger and more aggressive. This form of aggression by males is fundamental in the group form of society. (A group consists of many females and their pre-adult offspring, plus a dominant male. The male provides protection to the group as well as semen. There are many variations of this structure and some even include the presence of a limited number of non-dominant males.)

When Homo sapiens deviated from the pre-hominid branch of mammalian, there were many changes. As the female sexual availability changed from a few days of estrus per month to the whole month less a few days of menstruation, the group structure changed to the family (Family is one female plus one male plus their offspring.) and the society changed to the tribe. (Tribe is a collection of families, reflecting the advantage of numbers.)

However, in this transition the male brought along the genes that generate aggressiveness. The focus of aggressive behavior between males for the possession of a female remained but was generally limited to the pre-marital phase. (Again, the many exceptions do not invalidate the average behavior.) With the growth of different tribes and the general growth in population, there developed conflicts over terrain, of who had the use of the next valley. Aggressiveness found an outlet in the conflict between tribes that is the basis of war. (Is the Greek mythological story of Helen of Troy a reflection of this basic trend? Much of ancient stories have such a foundation.)

In a large society where war occurs infrequently, there is no acceptable outlet for aggressive behavior and apparently aimless aggressive acts often occur. There are many other factors that may contribute to such acts; economic frustration, abusive childhood, inability to attract a female partner, etc. This is a complex problem, the subject of many social sciences. The point being made here is that male aggressiveness has deep genetic roots that have only partially adapted to human societies.

9.4 Homosexuality

Homosexuality (gay, lesbian and bisexual) is genetic or hormonal based and gay or lesbian persons have no control over sexual preferences. Thus, we should no more criticize him/her for homosexuality than a Negro for his black skin or a hunchback for his abnormality. In addition, having happily "married" couples of the same sex should reduce the problems of overpopulation. This is difficult to accept by people brought up with the literal words of the Bible as absolute truths. However, there is a synagogue in Los Angeles that is openly homosexual. Many members are openly gay and the Rabbi is a lesbian. Oh, heterosexual people are also accepted as members. There is no discrimination.

This is a good example of how one religion, Judaism, changes to adapt to a changing environment. Some Jews are beginning to understand sexuality and are willing to bend the old rules. In a hundred years, the acceptance of homosexuals will be widespread. Again, in Judaism, consider the growth of reform, conservative and reconstructionism as attempts to adapt to the new environment established after the French Revolution. The Christian Reformation is an earlier example of slow and painful change of an established social entity. There were many subsequent divisions in the resultant Protestant Church. Although usually keyed to theological questions, or the egos of leaders, the underlying basis was adaptation to new conditions. These changes are good. The old Biblical rule, "Be fruitful, multiply and replenish the Earth" may have been applicable when the Earth was quite empty, but is no longer applicable when we are so overcrowded. Many organizations and individuals are currently studying the definition of a sustainable population density.

9.5 Celibacy

The voluntary non-practice of sexual intercourse by normal males is termed celibacy. It may be a genetically induced behavioral pattern that we have to accept as a variant from PL 2 as it does not hurt or harm others. As such, it is acceptable.

However, celibacy is also made a condition of priesthood in the Catholic Church. This has two benefits for the church. First, a male member of the church can allow the priest to work closely with his wife without any fear that the priest will attempt to have sex with her. Second, not having a spouse or offspring, all the attentions and efforts of the priest go to the benefit of the church. Any wealth that may have been acquired does likewise. The imposition of celibacy on a normal male is considered unSciEthical.

The celibacy forced on prisoners in jail is part of the punishment imposed on a criminal. In practice, this causes many problems of violence including the rape of young males and the practice of sodomy. The utilization of prostitutes in this situation should be considered.

9.7 Applications of PL 2

The urge for and the pleasures of sex were selected by evolution as they ensured the propagation of the specie. Thus, the sex ethics for many eons has been directed at enhancing the birth of children; then utilizing mother love and father love to ensure the growth, training and education of the child until it was able to take care of itself completely.

However, the development of modern society has lengthened the period of education to far beyond the starting of the urge for sex at puberty. Thus, in accord with PL 8, Adaptation to Changes, we will have to re-examine all the restrictions on sex described earlier.

9.7.1 Safe Sex

The experience of sex is SciEthical provided it is between consenting people, done in a manner to avoid pregnancy and to avoid the transmission of any disease. Being a virgin no longer has any value as DNA testing can establish biological parenthood. In addition, the exercise of sex by teenagers should in no way impede the completion of their education and training.

9.7.2. Pregnancy

Pregnancies are SciEthical only for a woman who is married and confident that her husband is able to and will take care of her and the child. It usually takes several years for a couple to become confident that their marriage is successful and will last. Thus, getting pregnant in the glow of the honeymoon while legal is a poor practice for either one. Getting pregnant while single in order to induce the male to marry her is also not SciEthical.

9.7.3 Rape

Rape—sex by force—is absolutely not SciEthical. The concept of rape has gradually become extended in our current society. In addition to forcible rape, it now also includes date rape where the male takes advantage by using special drugs, or the state of drunkenness. With the increased availability of jobs for women in a male dominated industry, women are often pressured to have sex without the use of physical force. This is also deemed rape and is not SciEthical. Essentially, rape is not involved only if the female is free to engage or avoid sexual intercourse without any risk of harm to her job or career. This also applies if it is the female boss who wants to have sex with a subordinate.

PART C

Chapter 10 PL 3—Marriage, Family and Bringing Up The Kids

10.1 Basic Purpose of Marriage

The basic reason for marriage is the need to protect and feed and train (educate today) the offspring in the evolutionary process that required more and more time for the child to develop and reach the maturity needed for self-preservation and procreation.

With evolutionary changes, the human fetus became larger and larger relative to the mother and she became less able to procure food for herself. The act of giving birth became more difficult and it often required the help of other women. After the child was born it needed to be fed, cleaned, protected for 2-3 years before it could take minimal care of itself; then it needs another 8-12 years to grow up and have enough strength and stamina to get its own food. It also needs about this time to become sexually active so that it can procreate the specie. These time periods are much longer than needed by any other mammal.

All this places a load on the female that is more than she can do effectively herself. A full time dedicated partner is needed to protect, feed and to help raise the kids. Thus, the joining of a single male and a single female in an intimate relationship called a family evolved as a solution to this problem of a larger fetus and longer childhood. The binding of the two adults is facilitated by the emotion of love. All these structures and emotions were selected by blind evolutionary forces.

Surprisingly, polygamy, the favorite social form in most mammals, is still an effective marriage system because it does provide support for the pregnant woman and for the child. The male who can support multiple wives must be rich or powerful. If the male's wealth is acquired by inheritance however, there is no guarantee that the grandfather's ability to acquire wealth or power is transmitted.

10.2 The Male Role

10.2.1 Earner

Whether he hunts wild game, fishes, gathers fruit, farms the land, keeps a herd of sheep, makes and sails boats or flies a Boeing 747, a man's basic role is to obtain the means to support himself and his family. He is the earner of material needs, whether he obtains them by physical, artistic or intellectual efforts.

10.2.2 The Strong Man

In the long development of mammalian species, the existence of two forms, male and female, provided evolution with many opportunities to select variants of each form that were better suited to the then existing environment. Thus, the structure, appearance, function, interests and habits of males and females grew differently.

In simplistic terms, the male became larger, stronger, more aggressive and focused more on factors outside the family. The female remained smaller than the male, more suited to bearing and rearing children and focused more on factors inside the family. This does not mean that the male ignored the family or that the female ignored the factors outside the family. There was considerable overlap between the two, but the centers of interest were not identical. This became a very efficient and prevalent difference between the sexes in almost all mammalian species. There are always exceptions! (Intelligence eventually proved to be more advantageous than brawn, and the brain size increased percentage wise more than the biceps.)

However, as the environments changed, both physically and socially, from nomadic small groups widely separated in a hunting and gathering economy to larger groups in a static farming economy, the expression of the aggressive trait in males started to lose its usefulness except in warfare.

With the advances of science in the last 300 years, machines have been substituted for manual strength and slowly the economic value of physical strength has diminished. Today the ability to push a button, flick a switch or rotate an internally self-powered wheel or lever is adequate for most jobs, although not all.

We have ennobled sports as an activity that values human strength, aggressiveness, speed and dexterity. Of course, many of us sublimate this genetic strait by just watching sports on the TV but youths, both male and female, engage in sports as part of the process of growing up. Thus, the widespread interest in sports is a sublimation of the male's aggressive spirit. The wife who complains that her husband is always watching football of basketball games on TV ought to be grateful he is doing that rather than taking his aggressive spirit out on others.

It is curious that while colleges educate and train our young people so that they can work in an environment that emphasizes intelligence, social skills and artistry, the same colleges place an overwhelming emphasize on sports that emphasize strength, speed and aggressiveness. Although there are competitive teams and tournaments in the fields of spelling, math and science, the recognition of accomplishments in intellectual competitions is almost lost in the publicity awarded sports.

The abuse of others by overly aggressive males is another example of the fact that many of our genes are more compatible with a life style a million years ago. The male genes have not yet fully adapted to the current generally peaceful, highly industrialized society where brawn, strength and aggressiveness are of less value and need to be sublimated by education and example. This is also another case of Nurture fighting Nature.

Aggressiveness is not totally useless. Intellectually, it is still used in the economic warfare between corporations, internal struggles within academia and within companies and in investment strategies to mention a few. In a certain sense, all progress is made by men or women aggressively pushing the existing limits of the social and physical environment. However, physical aggression against individuals is not now acceptable or useful in a SciEthic sense.

10.2.3 Model For Children

Another function of the male in the family is to provide a model of male behavior that educates the children. The girl learns what is expected of a male and what she needs to do to make him do what she wants done. The boy learns how to behave as a man and if the father is wise he teaches the boy how to work, when to fight, and when to run away. These are all instinctive impulses deeply rooted in the genes (Nature) but the cultural environment fixes the details of how the impulses are expressed (Nurture).

In households with a single mother plus small children, the absence of a male model is recognized as a potential source of misbehavior for sons and efforts are made to provide a relative or a teacher as a substitute male model. However, the girls also need a male model as well as the female model so one parent families with small children are basically unSciEthical.

Everyone can cite samples of single-parent households where the children grew up satisfactorily, but we need to reemphasize that social rules apply to the average and that exceptions will always occur. If the existence of an exception makes one want to disbelieve the rule, then this reflects the religious viewpoint that rules are made by a god and thus are absolute.

10.3 Female Roles

10.3.1 Caretaker

After the essential act of giving birth, the woman's chief role is that of taking care of the baby, nursing, feeding, cleaning and playing with it, which is really teaching it how to focus, move and to accept the existence of other people. This care-taking role has been extended to other people in need such as a disabled husband, elderly parents and eventually anyone in need of care. This role has changed markedly in the last 300 years as will be covered later.

10.3.2 Model For Children

Just as the male role includes acting as a model for both boys and girls, the female also acts as a model for both, but mainly for the girl. She teaches her the value of cleanliness, later making herself attractive to boys and the dangers of aggressive male behavior. In essence both parents, consciously and unconsciously, by example are teaching the children the skills needed for successful social interaction. Sometimes, but rarely, children are taught the economic skills of the parents and take over the family business. The technical change in society in the last 300 years has shifted the teaching of how to make a living from the family to the schools.

10.3.3 Surrogate Motherhood

With the advance in medical science, a woman can now accept ova fertilized in a glass dish and give birth to a child that does not contain any of her genes. This is basically an extension of the act of adopting an orphan and bringing it up as though it was her own. It is also an extension of the act of prostitution, using her body to satisfy someone else. In both cases, being paid for the bodily service is SciEthical.

10.3.4 Feminism

Following the developments of science and engineering, which substituted machines for physical strength, women have been doing more and more of the jobs usually performed by men. From working away from home in factories to providing care taking as nurses, women's range of paid jobs has expanded. In many cases, this also represented exploitation, as her wages were usually substantially less than that paid to men doing the same or similar jobs.

However, the rights of women as reflected by law and customs did not accompany this expansion of jobs. The fight by women to acquire these rights is called feminism. This fight was successful in that the right to vote, hold public office and receive legal protection for her special biological requirements has been obtained. The feminist fight continues against the glass ceiling and for the need of child care facilities on the job.

Like all trends, it also shows areas of over kill as women try to fill jobs as firemen and policemen, which still requires physical strength. In the military they can fly airplanes but service as infantrymen is questionable. As much as one favors unbiased treatment for women in all fields of endeavor, the fact remains that men and women are biologically and emotionally different and there are activities where this has to be recognized and accepted.

10.4 The Various Forms of Marriage and Family

10.4.1 Monogamy

The most widespread form of marriage among humans is one woman plus one man and the children they have conceived. This is called a family. Its widespread adaptation implies that the monogamous structure does the best job in raising children. Marriages may be legal or just factual, called common law marriages. Many families lose a parent due to death or divorce but manage to continue with the help of relatives or the government. There are many single-mother families that have never had a live-in father and that is a major social problem. In the slow process of evolutionary selectivity, children of complete families will do better.

10.4.2 Polygamy

Polygamy is the practice of a single male having several wives. This is common in Islamic countries, which constitute a large fraction of the world. To the extent that these families raise stable, competent and effective children, polygamy is an acceptable form of marriage. Cultural evolutionary forces do not make selections based on religious morals, but on the effectiveness of the practice. Thus, polygamy is SciEthical.

However, the fundamentalist Islamic practices also include complete domination of the females by the males and the forbidding of women to become educated and work outside of the family. Thus, they are reducing their productive potential by 50%. While not all Islamic people are fundamentalists, the latter appear to be dominating their countries.

It will take a long time for evolutionary social selectivity to show which form of marriage is superior. Polyandry, the form of one women with many concurrent husbands is rare, not growing, and will not be treated here.

10.4.3 Intermarriage

The growing global economy, immigration from Third World Nations to First World Nations, the widespread use of the English language in the aviation industry, science and the Internet are all symptoms of the One World envisioned by Wendel Wilkie in the 1950s. The existence of a growing body of international law and the activity of the United Nations support this picture. However, the multiculturalism of the 1990s in the USA is replacing the melting pot and supports a resistance to a common culture.

Nevertheless, from a long-term perspective, it appears that we are slowly and painfully approaching the One World structure. The creation of a single culture, a single race would help in reaching this goal peaceably. On this basis, intermarriage should be supported. This can be done by giving such marriages positive publicity, income tax benefits, and preferences in the education system to the offspring. Undoubtedly additional benefits can be created.

Although intermarriage between the descendants of immigrants from different European countries has become widespread and completely accepted, it is recognized that advocating racial intermarriage (although sizable) is still going against the current culture. The creation of a One Race will be more difficult then the creation of a One World.

While the marriage of two people of the same race who come from different cultures, i.e. Irish and French, is well accepted, the marriage of two people of different races, i.e. Caucasian and black, is tolerated but not encouraged. In the growth of an eventual commonalty of all people there can develop a common global race. From an evolutionary viewpoint, this removes the diversity, which permits selection of the characteristics best suited for the existing environment, which today is man made. From this viewpoint, intermarriage is not SciEthical. However, from the fact that human characteristics of all races overlap considerably, the selectivity of individuals marrying persons of another race may very well act to improve the average capabilities. From this viewpoint, and in consideration of PL 5, intermarriage is SciEthical. Only the future will show which way is better.

10.4.4 Extended vs. Nuclear Families

When the economic environment was agricultural, children stayed on the farm and the size of families became extended. This provided older people to look after small children and prepare meals while they were really being supported by their children. Handicapped people could also be taken care of and did not need to wander off and die. This was a very effective family structure.

With the advent of mechanical and intellectual industry, which were not tied to any specific land, people on farms started to take jobs in factories. With increased mobility, people could find better jobs further away. In times of economic depression, younger people left their homes and moved to places thought to be better. All of these factors broke up the extended family and eventually created the nuclear family: a father and mother and their children with no relatives nearby. This change was detrimental to the upbringing of the children especially when both parents need to work. Almost all people who remember the extended family regret its disappearance. However when the environment changes, be it nature or social or economic, we need to adapt. Evolution will eventually select children of those families that have effectively adapted to this change.

10.4.5 Divorce

We all make mistakes, even in choosing a partner in marriage. When one finally realizes the error in marrying that specific spouse, then one should be able to correct that mistake as one corrects other mistakes. The major problem here is that if there are children they will be damaged.

Thus, as mentioned earlier, one should not have children for a few years after marriage, but wait until one is more certain that the choice of spouse was correct.

An early recognition of this problem (Sanger, 1926) was the recommendation that two people live together for a few years, and if that is successful, then get married and have children. With the availability of modern medicine, this can be done without the female getting pregnant or any transmittal of a sexual disease. This practice is effective and thus SciEthical.

But sometimes the differences between the spouses develop slowly and the desire for divorce is not strong until there are children growing up.

At certain stages, a splitting up of parents can have a devastating effect on the child. Under this circumstance, the spouses should not get a formal divorce until the children have grown to the point that they can adapt to this change in their familial environment. Without a formal divorce, the adult spouses can make quiet and unofficial arrangements with other adults that can satisfy their emotional and sexual needs. This has to be acceptable for the sake of their children.

Basically, the ability to divorce a spouse is a fundamental right for every individual but it has to be done in a way that minimizes the damage to the children.

10.5 Applications of PL 3

10.5.1 Who Should Have More Children?

After staying alive and having safe sex, the most important SciEthic is having and bringing up children. In this period of over population, not having children would seem to be acceptable. However, if one is very intelligent, now the most important biological trait, it will improve the probability of the success of the social system if you have more children than do the far less intelligent members of your society. Thus, SciEthics says that if you are intelligent, have more children.

However, the trend today is just the reverse. Intelligent, well-educated couples with high incomes tend to have fewer children than those individuals or couples on welfare. The strain on society of this difference has become very apparent although rarely discussed. It is a "political incorrect" topic. Thus, SciEthics says to the intelligent persons, have more children even if it is counter to your economic goals and plans.

10.5.2 Extending the Family

The extended family being more efficient than the nuclear family, every effort should be made to have other relatives live nearby and be included in social affairs. When traveling, make every effort to visit relatives that may live where you are going. Retired couples tend to move to Florida or California in their sunset years, but your response should be to visit them there. Arrange your vacations so that you can visit relatives. In emergencies, visit and help relatives and if you have one, call on them. Do whatever you can to create strong family bonds and involve your children in such efforts.

10.5.3 Other Applicable SciEthics

There are many of these spread throughout the text but to avoid duplication, are not repeated here.

PART C

Chapter 11 PL 4—Membership In The Tribe Or Society

11.1 Introduction

Societal groupings predate and extend beyond Homo sapiens. Chimpanzees, apes, monkeys and lemurs gather in groups. Lions live in prides, wolves run in packs, deer browse in herds and birds fly in flocks. These social groupings all provide an improved capability for the individuals to live, procreate and protect their young until the cycle can be repeated. The detailed mechanisms may be different but the function is the same; improved survivability of the specie.

Small fish swim in schools. This provides the majority of them a chance to scatter and escape the larger fish that prey on them. In many herbivores, the older males form a picket fence around the females and the young and fight the predators off. The loss of a few males is not serious in a polygamous society. In addition, the stronger, smarter and more agile males that are left may very well improve the genetic makeup of the herd.

To the extent that social groups support the individual's safety and aid in getting food, they are "good" as his/her life is generally better than if he/she operated alone. In addition, an extended family, group or tribe will usually assist in the birth and rearing of the young and thus assist in the propagation of the specie. We will now focus on the growth of societies of Homo sapiens.

The growth of tribes into city-states, then states and now nations appear to behave as though the law of evolutionary selectivity is applicable. Many tribes and states die out and forms of government disappear. The concept of sovereignty is analogous to PL 5, The Pursuit of Happiness. Each sovereign state is free to do what it wishes within the constraints of its neighbor states. If a state is very powerful, it can apparently do whatever it wishes. International law, although growing, is weak and has no means of enforcement except through its members if they so wish.

The trend toward globalization has mainly taken an economic road, although The European Union is still an imminent possibility. The growth in size of organizations, be they economic or political, has an evolutionary aspect in that bigger is more powerful and thus better suited to survive as long as they can still adapt to changes. The same factors seem to apply to ethnic groups. Their diversity and multiplicity resemble species. Their ability to adapt is essential for survival.

Stronger cultures, European from 1400 to 1900 AD, drive out weaker cultures, i.e., American Indians, Australian aborigines. These do not completely disappear but exist only in small isolated groups. This is very similar to living species. "Might Makes Right" is still applicable although its process is somewhat less obvious today. Tribes and states decline and disappear, i.e., the biblical countries, then Egypt, Rome, Sparta, but their populations persisted, became smaller and adapted to the stronger invaders. The decline of the Roman Empire was a complex phenomenon resulting from internal changes more than from invaders. The abuse of power by its leaders played a major role. The nonperformance of the obligations of a state to its members and the corruption of the system destroyed the support of its people. Immigration resulting in dilution of a common heritage was a factor. In many cases, ecological changes caused by the unaware abuse of the natural resources contributed to the demise.

The growth and change of governmental forms from royalty to dictatorships and now to democracies reflects environmental-like selectivity. This is a theme that deserves much research.

11.2 Leadership

The greatly improved efficiency of individuals working together in groups preceded Homo sapiens and is evident in almost all species. The ability to defend, protect, find food and overcoming the loss of some individuals is overpowering. This is designated as PL 4. We have inherited that genetic tendency and strengthened it much beyond what is evident in simians. Although the advantage of group activity is great, it does not occur without some problems. The minor ones have to do with the fact that traits, preferences, and abilities are not the same in all humans but are distributed in some fashion, usually a normal curve. Thus, group activity often does not accommodate everyone.

The major problem however, is that a group activity requires a leader, someone to develop a strategy, assign roles and divide the benefits. The leader problem has been a major factor, both positive and negative, in social development.

The problem with leadership has to do with dividing the benefit. Any individual is prompted to get as much of the benefits as he can from any joint activity. If the individual is a leader, this provides him with the ability to take the best parts of the slain deer or other edible animal, more of the loot, and the best looking women captured in a raid. He usually does this, and the followers do not object as long as they all get significant benefits. They appreciate that they would not have gotten those benefits if the leader had not organized and led the raid.

The leader then becomes accustomed to taking a larger and larger share, favoring his family (nepotism) and friends (cronyism), thus reducing the amount that goes to the other members of the group. This can become so severe that the other members either leave the group (PL 9) or overturn the leadership (revolution).

Another failing of leadership is that when the leader becomes old or incapacitated, he turns the position of power over to his son or a close member of his family (An extension of PL 3), often regardless of that person's capabilities. Royalty did this for millennium. This abuse of power is today still widespread and deep in many societies. It can be destructive as seen in Indonesia and North Korea.

Selecting a new leader after a defined term in office is a method democracies use to avoid this problem. The advantages of this technique are much greater than the occasional loss of an effective leader who is retired too early or his growing ineffectiveness due to imminent retirement.

What are the requirements for becoming a leader and enjoying the extra compensation of that position? In early times, one had to be stronger and more aggressive than others in the group. Being taller and more symmetrical helped. Having a good idea (PL 6) as to a strategy and being able to vocalize it (PL 7) and convince others became more important. The effectiveness of the leader's strategies, appointments and activities solidified his position of power. If his effectiveness decreased or vanished, but the members of the group still received adequate benefits from the cooperative efforts, he remained in charge.

The requirements changed with time and the changes in the environment. Charisma is the term now used for the ability of a proto-leader to convince people to "vote" for him. Being in power, favoring and thus controlling the military and the police, and using it ruthlessly often keep a person in power even if his leadership is mostly ineffective. When the benefits for the average person diminish and disappear, that leader's days are numbered. The increased productivity of group cooperation must be distributed adequately so that the members would want to stay and continue to follow that leader.

11.3 Culture

Culture, as defined by Webster, is an all-inclusive term for all of the practices of a given society at a given time (disregarding the use of the word in agriculture and laboratories). Thus, the word "culture" as used here includes the language, government, laws, religion, manners, customs, arts and the behavior of its people.

To a large extent the culture of a society or its country indicates the state of its evolutionary development. If the culture includes human sacrifice (not punishment for a serious crime) and cannibalism, we term it primitive. If the culture includes the governmental care for all its individuals from the cradle to the grave, we term it advanced. These judgments, of course, simply reflect the beliefs of our current culture.

It is believed that the extent to which any given culture accepts and incorporates all of the PLs described in this book, will define and promote that culture's success and longevity.

11.4 The Social Contract

11.4.1 Introduction

The intent of the so-called "Social Contract" developed by Jean Jacques Rousseau in 1756 (Frankel 1951) between individuals and society is basically for the individual to:

    a. Defend it from its enemies through military services.
    b. Obey the laws of the society.
    c. Pay a fair share of its operating expenses through taxes.

In return, the society implicitly promises to:

    d. Protect the individual against harm from any source.
    e. Ensure his rights to his property.
    f. Ensure liberty and the freedom for his pursuit of happiness.

There is a tremendous spread of interpretations of each term in the above simplistic statements. There are many forms of society, governments and leadership and each will provide different definitions of the "Social Contract." Living in Iran and in Sweden and Southern California is living under radically different social contracts. But they all share a common aim, which is to define the rules of the game under which we play out our lives.

Additional rules are established by manners, customs, ethics and morals that may or may not be reflected in laws. These are parts of the culture of a specific society. It is believed that the manner in which all of the PLs are included and practiced defines a large part of the culture.

11.4.2 Society's Basic Duty To Its People

a. The earliest and still primary duty of a society, as reflected in its government, is to protect its people from forces outside the country. This is now being extended to protection of its people who are temporarily located outside the country. However, the ability of any government to protect its citizens in another country is very limited.

b. The secondary duty is to define the rights and responsibilities of the government and of its people via laws and the enforcement of such laws. Many of the rights and restrictions existent in a society are due to customs and manner and are not reflected in the laws. Many of the laws are not sufficiently enforced and thus their subject matter is also left dependent on customs and manners. The whole subject is treated in greater detail in the following section.

c. Enforcing the Laws: Justice Punishment and Compensation
A crime is a violation of a law written by the government of the society. No matter how heinous an action may be, if it doesn't violate a written law, as interpreted by a court, it is not legally a crime. For example, a man forcibly rapes a woman much against her wishes but not brutally enough to be judged battery. These facts are not in dispute. Normally, he would be convicted of rape. However, he is her husband and it is extremely difficult to convict a husband of raping his wife. Thus, the court decision would be different for the exact same set of physical actions. The court's excuse for these disparate actions is that there are mitigating or extenuating circumstances. Here the unusual circumstance is that there is no law covering the husband-wife sexual relationship.

d. In order to enforce its laws and induce others not to break them, society through its government exercises the power of punishment. Without punishment, laws would become meaningless. Actions against the well being of a given society as a whole, or that of individual members, would grow until the given society would disintegrate and its members disperse. Alternately, the violators could become the new government of the society and "enslave" or "free" the remaining members depending on one's group membership. In this case the initial society is gone, although the physical infrastructure and inhabitants may remain.

The unwillingness of the conflicting groups in Lebanon to enforce or obey the laws of a given regime led to crimes of terrorism, economic piracy and outright slaughter of members of the other groups. Today, Lebanon is a disintegrated society. Thus, the ability to punish and inhibit undesirable activity is considered a necessary characteristic of a government.

The punishment in today's society is usually directed at the criminal and in the form of loss of wealth, i.e., fines or confiscation of assets; loss of liberty, i.e., confinement in prison or restricted activity while on parole; or loss of life. Physical punishment by whipping or loss of body parts is no longer used in Western society, but still prevails in some Muslim states. In all cases, the emphasis is on punishing the lawbreaker.

If one considers that the intent of the so-called "Social Contract" between individuals and society is basically for the individual to obey the laws of the society, pay a fair share of its operating expenses through taxes and defend it from its enemies through military services. In turn, the society implicitly promises to protect the individual against harm and to ensure his rights to his "life, liberty and pursuit of happiness (and property)." There is a tremendous spread of interpretations of each term in the above simplistic statements. There are many forms of society, governments and leadership and each will provide different definitions of the "Social Contract." Living in Iran and in Sweden or Southern California is living under radically different social contracts. But they all share a common aim, which is to define the rules of the game under which we play out our lives. What are the rules for crime and punishment in a society living by SciEthics?

Although the laws in a SciEthical society would be somewhat different than in a religious society, a crime would still be the same, i.e., breaking a law. The nature of the punishment would be somewhat different. Although still directed at the lawbreaker, the emphasis would be on providing the restitution to the victim. In a sense, the lawbreaker and the society have not lived up to their contracts. The lawbreaker did not obey the law and society did not protect the victim. His suffering should be alleviated. This is not a morally derived conclusion. It is derived from the fact that if society does not uphold its end of the bargain, members will leave and go elsewhere or form a new society. This will be the end or death of that society, In evolutionary terms, that specific specimen society will not live and certainly not propagate and eventually the society specie will become extinct.

Consider the Divine Right of Kings and 18th century France. The remnants of royalty that now remain do so only by giving up their power and becoming symbolic. Thus, the societies and/or governments that have not died are those that have more or less held up their part of the contract, at least to the satisfaction of the larger part of their populations. Where people are able to vote with their feet, they leave an unacceptable government. The need for a Berlin Wall illustrates this. When Hungary tore down its barbed wire wall with Austria, Eastern Germans started flooding Hungary in order to get to Austria. (People may leave for other reasons as well, i.e., economic, drought, to rejoin families, etc.) Thus, the living and propagating form of society is one that keeps its bargain to an acceptable extent.

Thus, the thrust of punishment should be to undo the damage to the victim. For crimes that involve stealing of or damage to assets, the lawbreaker should be compelled to recompense the victim, be it an individual, corporation or the state, whose assets were taken or damaged. The funds for the compensation would come from the wages earned be it in or out of jail. Jails should be factories, farms, construction sites, forestry camps, etc., as well as liberty-restricted facilities. This labor should not be forced. But no work, no food. This situation is true in our society outside of jail so why not inside? He is being provided with a job, food, clothing and housing. This is more than current society ensures its law-abiding citizens. If the inmate refuses to work and starves to death, so be it. He has not adapted to his societal environment. The dictate of evolution is "Adapt or Die." If inhabitants of a society do not want to obey its laws, they should be free to advocate changes in the laws or in the government or to leave for another society. But not to violate the laws or harm their fellow citizens.

Salaries in the penitentiaries should be appropriate to the same labor provided outside, and the facility should be operated on a profit making, free enterprise basis, but with the freedom restriction costs borne by the state. Some equitable distribution of the inmate's wages between room and board and victim restitution would need to be made. These proposed jail/factories can include any activity compatible with the security requirements of imprisonment.

Various modes of punishment can be visualized. Parole with a job, working outside but being confined nights and/or weekends, etc. But the thrust should be restitution to the victim, not just punishment to the violator. The ability of unskilled workers being able to earn enough to make restitution and the length of time required remains a problem. Since the state also failed to protect the victim it would not be unreasonable to have the state reimburse the victim promptly and have the prisoner pay off this debt to the state over time.

e. If the restitution just equaled the damage, then the potential criminal is handed a no-lose situation. The restitution should be some multiple of the loss. This multiplier factor should be high for wealthy white-collar criminals who could avoid any punishment by paying all damages. In a complex society the rules of the game will also become complex. Lawyers would have a great time defining the rules under SciEthics.

The difficulty of the restitution approach increases when the criminal maims or kills someone. What is the compensation assigned to some one severely and permanently injured, i.e., made a paraplegic by a drunken driver? Or to a wife and family where the father and bread- winner is slain? There is no way to compensate in kind or in value for this type of loss. However, let us consider that the same loss can occur by accident for which no one is responsible or even when the victim is responsible, i.e., the drunken driver breaks his back in the auto accident and now has a wife and children without support. Or the father becomes ill and dies of natural causes. What do the innocent victims do in such an event? Widespread insurance or state responsibility would only be able to provide the necessary monetary aid. Thus, having the person responsible for the human loss pay the monetary aid in the end is no different as far as the victim is concerned. All the victim can get in the injury or loss of a loved one is the income that the loved one would have been able to provide, plus the direct costs due to the event. The only option we have is who provides the lost income, all the people through taxes or the party responsible for the loss?

What should be done if the person maimed or killed is himself responsible for the event? What would have happened in more primitive times when the effects of evolutionary forces were not compensated by accident insurance or social security? Should we still allow evolutionary selectivity to operate? Do we want people who are stupid enough to drink and drive or snort cocaine and walk out of a window to leave any offspring?

f. With regard to punishment, how does the process presented above act to prevent crime? Should we not put executions on TV and in the program explain the criminal act that led to the execution? Should we not then encourage parents to have their children watch that TV show?

There are many situations where the deterrent effect of possible punishment is ineffective. For example:

Crimes of passion—Blind impulse without any plan or intent.
Extreme poverty—One has little to lose.
Gain vs. risk—A gambler miscalculates the odds.
Political idealism—Assassinate the enemy of the people.
Revenge—Getting even for an imagined or even a real injury.
Religious fanaticism—Killing a doctor that performs abortions.

The thought of punishment may not be effective in these and other cases. Prevention through education is probably the best approach. But this begs the question of what we do now? Should one again consider the question of evolutionary selectivity? Do these actions represent an incompatibility with society? If so, does it reflect inadequate education or a genetic trait? At this time we do not know, but with the continued development of the Genome we eventually will.

11.4.3 The Growth of Responsibility of Society

Cooperate With Your Neighbor. Cooperate with your neighbor normally means working with him to achieve an objective that the neighbor could not do by himself. Two or more people working in unison can accomplish what is impossible for the same number to try to do individually. The classical tribal examples are hunting game and warfare. A more recent example is raising a roof over the shell of a structure your neighbor is building. One does these things willingly because of the greater benefit to oneself as in hunting, greater security as in warfare, or in the understanding that if you help your neighbor raise the roof of his house he will later do the same for you. The mutual benefits understanding (reciprocity) has grown or expanded so that your neighbor will help you to dig a well rather than the exact repayment of your help by only helping raise the roof over a structure you are building. How far should/could expansion of this tacit agreement go?

The principal of reciprocity also includes "do him no harm." (Note 8) Like many of the Ten Commandments, it is a restriction on one's actions, a negative ethic. It has, in practice, been extended to a positive one of helping a neighbor in need. If he falls and breaks a leg, you carry him home and depending on the state of medical knowledge either amputate the leg or straighten it and put it in a splint or cast. Again, this is potentially mutually beneficial because if you broke a leg, you would want him to do the same for you. This extension has grown to where if anyone, not even a neighbor, or not even someone from your own tribe or coloration, has an accident, most people will quickly help or at least call 911.

Is there a limit to this extension of PL 4? (The above example is really also an extension of PL 1 as well; the accident victim might die if help is not quickly provided) If we are responsible for our neighbor's life in the event of an accident, then are we responsible for his having food, shelter, clothing and medical attention when there is no accident? This responsibility, from cradle to grave which exists today in some Scandinavian countries, also includes keeping him alive as long as possible within the abilities of the current medical structure. If he has a family, then this support would have to include them as well.

Thus, the responsibilities implicit in the reciprocity concept of PL 4 are increasing and becoming the responsibility of society. This just means that even if we are not personally involved, we pay the taxes that permit others to be. Where does this responsibility stop? What if the neighbor, or any member of our society, is responsible for his own inability to support himself or his family? What if he is alcoholic or a drug addict, cannot find a job, or is just plain lazy and incompetent? However, what if he is not responsible for his own inability to support himself and his family? What if he has Downs Syndrome, cerebral palsy or an IQ of 50 for a variety of medical or genetic reasons? What if he has had an accident that resulted in becoming a paraplegic? Would it make a difference if the accident was his own fault or if it was not?

Going back to the beginning of Homo sapiens, the evolutionary process selected those biological characteristics that enhanced the survival of the specie. Among the most significant ones was intelligence, creativity, the ability to talk, curiosity and the urge to experiment. Since Homo sapiens inherited a group culture from their simian ancestors, this evolved to a tribal culture. There is cross-linking between biological characteristics and societal characteristics. Observance of PL 4 helped the tribe to exist and grow so that individuals with a strong feeling of cooperating with and not harming their neighbors were biologically selected. This is a cross linkage between Nature and Nurture. Individuals with certain biological tendencies, such as helping others, that enhanced tribal success ensured the selection and enhancement of those tendencies. This is believed to be the evolutionary force behind the extension of the first simple PLs 1-4 to the current school of political thinking that the state has the responsibility for supporting all individuals from the cradle to the grave. Of course, not all individuals will agree with this.

The evolutionary force that selects those characteristics that enhance specie survival also allows the demise of individuals who do not have those characteristics. The term "enhanced survival" also implies enhanced non-survival of individuals lacking those successful characteristics. Evolutionary change is the sum of pluses and minuses.

There is a counter argument that those individuals not having the desirable characteristics would have continued to successfully survive as they were doing before the mutation/change in one or more individuals that created the start of a new or enhanced desirable characteristic. This counter argument implies a static environment, which if it had existed, would permit the continuation of the specie in the form it originally had before the change or enhancement. However, the environment is usually always changing. Most species, 98% of those identified, have vanished. In addition, the individuals with enhanced characteristic are "different" and would separate and form their own group or take over the existing group. In conflicts between groups, they would win and accelerate the disappearance of the older variety. Thus, the new or enhanced characteristic itself creates a new environment that although societal rather than natural, is nevertheless a new environment.

If we explore the environmental selection minuses, it is apparent that many of the deficient individuals kept alive today by medical science and/or by the cradle to grave security provided by society would have perished earlier in their lives in the past. Thus, to some extent science and society today are suspending the evolutionary process of minus selectivity. Whether science and society are also suspending the evolutionary process of plus selectivity is a question that will be skipped for now and here.

If we accept the fact that Homo sapiens is here now because of the process of evolutionary selectivity, both pluses and minuses, then suspending that process will prevent the appearance of Homo superiors and may allow the disappearance of Homo sapiens. It is not a decision we can make. If the changes or enhancements are sufficiently strong, it will happen anyway. Undoubtedly we can influence the process biologically through genetic selection and engineering and socially through special schools for bright students rather than for deficient students. The topic of deliberately enhancing the shift to Home Superiors will be covered in Part D.

11.4.4 Citizen's Duty to Society

a. The primary duty of the citizen is to fight the enemies of the country. Thus, participation in the armed services is a required duty of its inhabitants. Exceptions may be made for people with extreme disabilities, mothers of young children and others.

However, exempting women as such is no longer applicable. Science and engineering have removed the need for strength and speed. Women can drive tanks and airplanes, punch buttons and aim cannons. (Can they load them?) Making duty in the armed services voluntary is also inappropriate. Making everyone spend a limited period, say two years, should be adequate for training purposes. It would also provide all inhabitants with a common experience. Pure volunteerism is making the services a refuge for people unable to make it in our economic society and creating diversity rather than unity.

b. The secondary duty of the citizen is to obey the laws and ethics of the society. This ensures that all other members of the society will be safe and able to pursue their individual paths to happiness. If all the other members also obey the laws and ethics, then the citizen will be able to do the same. Does this formulation beg the question? Let's see.

c. The next duty of the citizen is to participate in the formulation of the laws so as to ensure that they truly reflect the scientific basis of ethics, as he understands them in terms of the data available and his education. As social biological research continues, it will provide more data. This data may then lead to new or revised hypothesizes and theories. Both will affect the education and the understanding of all citizens. In the light of these new data and theories, it may be necessary to change or modify the laws and ethics in order to make them more effective in providing each citizen with improved opportunities for "Life, Liberty, and The Pursuit of Happiness."

d. In addition, as research and development continues in the other fields of science, new situations may arise, which are not covered by the existing laws and ethics. Technology has been the driving force in changing society. Thus, we will have both a changed understanding of how we should handle old situations and the development of new situations for which there is no precedent. The question then arises as to the qualifications a citizen must have to participate in the formulation of laws or in the selection of representatives to do that. Does a non-educated, stupid and bigoted citizen have the same voting rights as a college graduate with a high IQ? This question has a deja vu sound from the arguments that went on at the start of our Constitution in the 1776-89 era. Whatever we have learned in the two centuries since then should be reexamined, studied and the objective conclusions applied.

11.5 Applications of PL 4

Let us now examine the application of SciEthics to some of the questions arising in the extension of the implied intent of PL. 1, 3 and 4.

11.5.1 Life

PL 1 states that staying alive until you have progeny is an absolute necessity if the specie is to survive. Because of the length of time children need to grow, parents must be there to feed, protect and train (educate) them until they can become self-supporting. This is the basis for the family and PL. 3. If we approximate the age of self-support and marriage in a primitive society as 15-18 years then the parent must live to 30-36 before his/her existence is no longer vital. If one adds to this an insurance policy for taking care of the grandchildren if their parents are killed, then this becomes 45-54. Of course, their availability as baby sitters allows the parents greater flexibility for projects requiring trips.

In addition to the initial couple's availability to ensure continuation of the specie (PL 3), there is the fact that a person who has lived to 45-54 has acquired a lot of experience and was a valuable asset to the community for the millions of years before written records were invented. Oral tradition reflects this experience. Having someone around who could remember what happens during an earthquake, volcano or forest fire, and what to do in each case is an insurance policy of a different kind. Thus, as long as the adult remembers and is competent, able to take care of themselves, he/she is an asset to the community.

Although a life of 45-54 is adequate for perpetuation of the specie, it represents the minimum goal for the bulk of a population. Allowing for an extra generation, it would appear that living to 60-72 should be completely adequate. However, due to the distribution of human characteristics, some people can live much longer, say 90-100, occasionally 110-120. This feature is sometimes described as "over kill" but really represents the small probabilities associated with the upper end of any statistical distribution. Of course, many people also have shorter lives than 45-54.

However, if and when the adult has lost their memory, becomes bed ridden or comatose with no possibility of recovery, they can no longer be of service to the community and thus represent a drain on its resources. Their continued life is a minus from an evolutionary viewpoint. They should be allowed to commit suicide or be helped to do so. Because of PL 1, this should not be imposed on a person against their will. It should be an available alternative course of action. In primitive societies with limited resources, the people roamed the wilderness looking for food. Older people, unable to travel on foot with the rest of the tribe would slowly walk away into the forest or desert to die in order to avoid being a drag on their family group. No one stopped them. It was a part of their primitive culture. In our current culture, this act is often done surreptitiously.

The justification of forbidding the deliberate self-ending of a human life and thus terminating the expenditure of limited resources rests on the religious precepts that life is divine. God has given the body its soul and only God can take it away. The attractiveness of this religious premise rests on the fact that the will to live is so strongly genetically built into the psyche of the individual that he/she never wants to die. It is another over kill. The concept of immortality is emotionally very attractive.

It is easy to point out cases that do not follow the above conclusions. There are always exceptions to any rule, except perhaps in mathematics and physics. As discussed elsewhere, all human characteristics are statistically distributed in some way; the normal curve is the most widely known way. However, the conclusions reached from SciEthics are generalizations that apply to the majority of mankind.

An interesting observation is that a fair number of elderly people when they become ill with little hope of cure, will want their estates or savings to enhance their children's lives, rather than be wasted on giving them a few more months or even years of unpleasant life. This is a strong example of an evolutionary plus characteristic.

Competent younger people with serious illnesses, in great pain, also often want assistance in committing suicide. In accordance with PL 5, this should be permitted. From the societal viewpoint, funds expended in keeping them alive for a few years is a waste of resources better spent elsewhere.

From a SciEthic viewpoint, life is essential up to a point. After that point, it may represent a waste of resources and an evolutionary minus. Determination of just where that point is and how to prevent abuse of the ability to end lives represents detail problems that should be explored.

11.5.2 Living

Is society or the state responsible for providing food, shelter, clothing and medical attention for everyone in the territory of that state? Let us rephrase that question as, "Does providing everyone in a given society or state with food, shelter, clothing and medical care for their entire life continue or hinder the progress of Homo sapiens to Homo superiors?"

The reformulation of this primary question constitutes the substitution of continued enhancement of the specie instead of ensuring the sanctity of life as the objective of society. On this basis, the evolutionary process of pluses and minuses should be allowed to continue. A secondary question then becomes, "Do we have the wisdom to know what to do to allow this when the environment in which we all live becomes more and more created by man and not by nature?" Nature is amoral and blind and has no goals. Man thinks he is moral and has goals; many of which are conflicting. Because man is intelligent and his knowledge is increasing, we are filled with hope and will assume he can do the right thing. However, the shift from theological goals to the goal of supporting the development of Homo superiors will not be easy.

Returning to the reformulated question, society should allow the painful minuses of evolution to remain effective. This means that irresponsible individuals who cannot support themselves in our society should not be perpetuated. If this idea is neglected, then this society will not grow as fast as it could or last indefinitely. Another way of exploring this problem is to conduct an experiment and allow multiple societies to try different degrees of support and let "Nature take its course." Unfortunately, there are so many forces at play in any society that this is not a simple, clean experiment. Also, the people in the severe treatment society will emigrate to a better treatment society, legally or illegally, thus negating the experiment; or if the better treatment society grows and prospers even with this immigration would that prove its better treatment is superior? Or would other factors predominate? Is it possible to set up an experiment that would provide firm conclusions?

As an example of the application of this approach in our society, let us make the use of hallucinogenic drugs and alcohol legal while teaching everyone of their dangers. Anyone stupid enough to take to them will suffer and without government support will have fewer children or will not train/educate the ones they do have in a manner that will perpetuate their genes. Thus, the average intelligence of the population should increase. This will be cruel and hard on many families. In any case, minors, who are not responsible for their parents' behavior, will have to be taken care of by the state. This is admittedly a very simplistic approach. With the advances in knowledge from the Genome Project, it will probably become easier to relate cause and effect and define more effective tactics. Recent discoveries have shown that there is a strong genetic influence on personal behavior.

11.5.3 The Minuses of Evolution

Implicit in the approach of letting the minuses of the evolutionary selective process continue to operate is that the support of people would be limited and that more effort would be expended in training or education so that these people can operate in the current economic environment and become self-supporting. There is no acceptable way that they can be made to vanish. (PL 1 is still No. 1) A corollary is that the economic system must be structured so that there are jobs for everyone who wants to work, even people of limited capabilities. (A multi-tiered society is inevitable. Even when the proletarians were successful and become the dictators in a society, a multi-tiered structure developed.) The continued replacement of low-skilled people by machines and computers needs to be reexamined. Opportunities for them in public works projects should be considered. A corollary problem of course is population control. This involves control of immigration and birth rates; each of which is a major problem. In view of PL 2, it would seem that research on and the application of methods of birth control should have high priority rather than be verboten.

What to do with the people who are unwilling to live by these standards and who prey on their neighbors is a problem we will bypass here. In science, one works on problems for which there is an approach. The isolation of such people in work camps and prisons is one approach, and then trading time for permanent sterilization may be effective. Originally, the forest was full of tigers, now the streets are full of muggers. The helical theory of history appears to have some merit. Perhaps we just have to be as careful in the street as our ancestors were in the jungle. Is carelessness only an evolutionary minus which we have to live with? Or can it be prevented by education and training? Nature versus Nurture again.

PART C

Chapter 12 PL 5—The Pursuit of Happiness

12.1 The Freedom of Choice

The freedom of choice in the pursuit of happiness implies that you are free to do whatever your heart or instincts prompt you to do. The desire to stay alive, have sex, have a family and be a member of a tribe have been covered in PLs 1-4. This section is concerned with a person's activities in the context of a group or tribe or society.

You are free to select any economic activity; hunt, fish, drive a truck, be a salesperson, become a singer, actor, dancer, scientist, inventor or whatever your heart desires, and you are capable of doing. There may be an intrafamily conflict, as parents often want to tell their children what to do from working on the family farm or in their father's business. This conflict is usually resolved by the young person leaving the family and going their own way.

If what you want to do is counter to the culture or laws of your society then you are free to leave that society and immigrate to another region where you can do what you want to do. This right of emigration does not include the right of immigration if there are people in that other region who object to your presence. You then have the option of becoming an illegal immigrant or going elsewhere and finding an empty valley and starting your own tribe. People who leave their own country usually take their bride along or have her join him after becoming established.

This freedom to pursue your own road to happiness creates a social variability, which permits evolutionary selection in society in a manner analogous to evolutionary biological genetic selection.

12.2 Restrictions

There are certain restrictions, of course. Your actions should not hurt nor harm your neighbor. (See Note 8 for definitions.) They should not be illegal except under extraordinary circumstances.

Doing or saying something that offends your neighbor is permissible. Offending what other people think is the basis of change. If we couldn't offend anyone else, how would we change our culture, i.e. our regard of gays, lesbians, women, people of different color, physically handicapped people, etc. The fine line is between the freedom of speech (PL 7) and saying things that induce some people to do hurt or do harm to other individuals. The Supreme Court has ruled that you cannot yell "FIRE" in a crowded theater. As an extension, you should not advocate that if you feel blacks or Jews are inferior then kill them. There are many things we can do that only offend some people. Let's explore two of these things next.

12.3 Applications of PL 5

12.3.1 Nudity

A nude person walking down the street does not hurt or harm anyone else. It does offend many people, and this is reflected in laws and customs, making nudity illegal. There are many holes in these laws and customs; nude beaches, strip teasing theatricals, partially or totally nude waitresses, small children up to an ill-defined age, etc.

There is a danger to a nude woman in the above circumstances, that her nudity might be interpreted as an appeal for sex and thus induce a rape. In primitive tropical areas, the custom is for women to go bare breasted. This does not create any problems. Even in the primitive tropical areas, men wear loincloths and nothing else.

Are there reasons for the preponderance of clothing in most cultures other than the weather or decorations?

12.3.2 Other Religions

Many sects are variations of generally accepted religions. Here we will focus on one that is usually offensive. If one believes in Satan and worships him, that's all right as long as there are no human sacrifices. The right to sacrifice animals, as a substitute for humans was a tremendous improvement, initiated by Abraham in order to save his son, Isaac. It is not SciEthical, as it harms the economy and does no good. Torturing animals is also not acceptable as it creates the danger that participants will slide down a slippery slope and extend such treatments to humans.

In general, the practice of any religion is acceptable as long as it does not do, or induce others to do, any hurt or harm to others. Is Sadism an exception?

PART C

Chapter 13 PL 6—Intelligence

1.1 Introduction

As pointed out previously, the growth of intelligence was one of the most important factors in the increased ability of Homo sapiens to survive, increase its numbers and be able to overcome all the other carnivore animals that were bigger, faster and "meaner." Starting with the domestication of animals, the cultivation of plants, industry and finally science, mankind demonstrated its superior power and capabilities. We have now put a man on the moon and are planning to do the same on Mars. The prospect of eliminating diseases and of genetically engineering Homo superiors looms in our far future. (More about this in Part D.). All of this is due to brains and not brawn. Another facet of this superiority was the development of larger societies with complex structures. Enhanced by larger numbers, these social groups could do what the individuals or smaller groups could not. This has been discussed in greater detail in Chapter 11, PL 4.

13.2 The Growth of Applications of Intelligence

Initially, much of the benefits of inventiveness and creativity was to the individual, then family, and finally the group or tribe. This must have included tools, clothing, pottery, housing and weapons.

However, much of the subsequent creativity was beneficial to the social relationships within the tribe, such as music, dancing, singing and rules of behavior. Ingenious explanations of why Nature was the way it was led to inventing the concepts of gods which led to organized religion.

Another type of inventiveness was the substitution of games for the acts of warfare that allowed males to establish superiority without slaughtering their opponents. Of course, this occurred slowly; gladiators in ancient Rome, duels in the Middle Ages, and fist fighting in the 20th century persists 2,500 years after the Olympics. But today in general, sports are widely used as a means of satisfying an ancient innate genetic tendency or drive for superiority without serious damage to the individuals involved. These changes arrive slowly and overlap in time. This is progress in an evolutionary sense.

13.3 The Misapplications of Intelligence

A problem arises when persons use their intelligence to commit crimes and hurt/harm others. This behavior reflects the influence of Nature and Nurture, of genetics and upbringing. Our genes produce traits or tendencies that are modified or expressed in different terms of the culture one is raised in.

Then some of our genes go back to pre-human or hominid species and even to mammalian species. Thus, we may carry in our genes traits that preceded our current species. Many of these traits have became obsolete as Homo sapiens developed. Individual human behavior can be triggered by genes that are obsolete or missing for the majority of people, but still exist in some individuals due to genetic variability and the slow extinction of no longer useful genes.

13.4 The Multiplicity of Intelligence

The word "intelligence" has acquired a specific meaning described by an IQ number, the results of a written test. This is 100 for the average person, 140 and above for geniuses and 60 or below for mentally handicapped people. The IQ number of many people are distributed along a normal curve. The nature of the test has been criticized as being culturally biased and unfair to minorities that have not had the benefit of a full education.

The existence of differences in IQ means between different ethnic groups (Herrnstein & Murray 1994) has also been questioned, although there is a marked superiority of certain Asians and Jews. We will not get involved in this argument. Instead, we will cover the misapplication of the word "intelligence."

The growth of humankind reflects the contributions in many areas of creativity. From tools to metals to airplanes, from cave wall painting to massive sculpture, from simple hymns or songs to opera, from jiggling in unison to modern dancing, from oral stories to the Encyclopedia Britannic; all these have contributed. The essence of mankind is that he could, and did, continuously change his modus operandi. To the best of my knowledge, no animal has done this. Except possibly the beaver, which builds dams and homes. Birds build nests and insects build hives. But we have not observed continuous change in any other specie except for animals living in parks or mountains near human residences. They scavenge from food left outside, garbage pails and from eating small pets and small children.

Thus the term "continuous creativity" or creativity for short would be a better term than intelligence in describing the human trait that has led us to be the most powerful living form of life on this planet. The subject of emotional intelligence has received a lot of attention lately, but it can be considered as another form of creativity. Emotional intelligence must have played a major role in the growth of groups and tribes into states and nations.

Although the word "intelligence" is used throughout this book because it has such an overwhelming current usage, it is really used herein as a substitute for creativity.

13.5 Applications of PL 6

13.5.1 Social Valuation

Society should put greater emphasis on the achievements of intelligence than on the achievements of brawn and speed. The emphasis on sports in the news is astonishing. The publicity about students in science fairs and in science competitions is very small. It is unfortunate that the fraction of the population that can understand sports but not science is so large.

13.5.2 Government Spending

In view of the success of scientific accomplishments that have resulted in the advancements of society, more funds should be spent on research. Cities should spend more money on education and avoid spending money on new sport arenas.

13.5.3 Education

The recognition of intelligent students should start early and be sponsored and advanced. The funds so used would not be large because their number is much smaller than the students of limited ability. But emphasizing the support of intelligent students, talented in any creative way, puts creativity in the spotlight where it will attract other students. Let's invest in what advances society.

PART C

Chapter l4 PL 7—Free Speech

14.1 The Development of Speech

Somewhere and sometime in the evolution of Homo sapiens, the ability to grunt and snarl grew into the ability to speak. Evidently, the ability to communicate verbally was extremely advantages in order for it to be selected. This was a Nature selectivity, as it required biological changes in the larynx and vocal cords. It is surprising that no other mammals have acquired this capacity, although it is not yet clear whether the sounds of whales are speech or complex grunts or something in between.

14.2 Gains From Communications

The gains from speech as an improved method of communications are enormous. The ability of a leader to assign different tasks to members of a group is essential for the group activity to be coordinated. The ability to pass on experiences and the descriptions of enemies are also essential for the development of a society. The teaching of abstract concepts of right and wrong to children growing up would be limited. Learning would be somewhat limited to copying what the adults did although snarls would inform a cub that what she was doing was wrong.

The development of philosophy, literature, sciences and most arts would not have been possible without speech. Speech provides a feedback mechanism for people to inform their leaders that what they were doing was good or bad. Negative feedback is a requirement for stability of any system, mechanical or social.

14.3 Applications of PL 7

14.3.1

Give others the benefits of your thoughts, i.e., do not be silent when you see others do something that is unSciEthical or are apparently undecided what to do. This is part of the process of passing on wisdom to others or teaching young people what is the right thing to do.

14.3.2

Allow others to do the same, i.e., freedom to speak applies to every one else although with some restrictions.

14.3.3

Tell the truth although white lies are an exception. White lies are lies that do not cause anyone harm, i.e., telling a dying man that his son is on a trip when actually he is in jail.

14.3.4

Do not reveal material told you in confidence and which you accepted as such.

14.3.5.

Do not make statements that induce others to hurt or harm some one.

PART C

Chapter 15 PL 8—Adaptation To Change

15.1 Major Changes In The Physical Environment

If our ancestors had not been able to adapt to changes in the environment, then as the dinosaurs, we would not be here after the Ice Age. Since there are usually many apparent ways to adapt to any change, the ability of individuals to pick different ways provided nature with various alternatives from which the best ones would eventually be selected. Thus, the freedom of an individual to take what action he/she thinks best is safest from an evolutionary process and becomes PL 5, which is equivalent to "Liberty" in the "inalienable right to Life, Liberty and the Pursuit of Happiness" clause in our Declaration Of Independence, mentioned before as a philosophic ethic.

The ability of the Homo sapiens to adapt to physical change apparently started with the shift from arboreal habitat to walking the savannas. It later became the ability to use fire and the skins of furry animals as clothes and move into caves or build wooden equivalents when the Ice Age arrived. Some must have elected to move south, which would provide the benefits of diffusion of the specie. Again, the ability to select different solutions to the problems arising from the arrival of the Ice Age gives the evolution process selection options. The ability of Eskimos to live in the far north is an example of extreme adaptability.

Subsequent to the Ice Age, changes in humans' environment resulted mainly from technology advancements, starting with the development of agriculture and the use of domesticated animals. It is difficult to separate the effects of technology as changes in the physical environment or in the societal environment. For example, is the change from a hunting/gathering-wandering behavior to that of a settled farmer and domestic animal herdsman a change in physical environment or in societal environment—or both? In many ways, the technological advancements force changes in both environments.

15.2 Major Changes In The Social Environment

Major changes in the social environment occurred as technology changed wandering tribe of hunters to settled farmers and herdsmen. The increased productivity of food and the reduced risk of hunting allowed the growth of activities by creative individuals who now did not need to hunt or grow food. Also, the leaders of the larger and more concentrated population had many more new problems to worry about, which led to more "laws" and larger governments.

As the tribes grew larger they occupied all the empty valleys, but remained as a single ethnic society or tribe. There soon were boundaries between diverse ethnicities and warfare developed for various reasons. But for this chapter, let us look at the adaptability problems within the changing societal environment of a single society.

The most outstanding social change was the growth in power of the "leader" of the tribe that grew into a society and the consequent development of "governments."

With the increased food productivity of the agricultural technology, both farming and herding (and later fabrication), the leaders were able to establish systems of taxing the farmers and fabricators in order to provide funds for their own use, establish laws and a judiciary in order to avoid the need to hear all complaints and a permanent army to ensure both the safety of his country and to ensure his own reign. In addition, he usually established a group of loyal supporters, or noblemen, by giving them administrative power over small sections of his territory. Thus, local natives had to obey both his noble and the leader of all the nobles. The nobles formed mini-states in today's nomenclature. The individual farmers, tied to the land in many countries became slaves or "serfs." The fabricators, not being tied to the land, remained somewhat freer to move around. The original alternative of moving to the next valley had disappeared. Power is indeed corrosive and is abused.

The common people adapted to the impositions forced on them by a powerful monarch supported by his army and armies of his nobles. It is better to be a slave or a serf that to be dead, according to PL 1. The monarchs and nobles always allowed sex and marriage in order to ensure themselves of more slaves or serfs. Thus, PL 2 and PL 3 were observed by the government. Generally, the serfs lived in communities so that PL 4 was also observed, but with limitations.

In parallel there emerged the more formal organizations of religion, which also "tithed" its members for financial support. Usually, the reigning monarch and the leader of the church cooperated after the religion gained a considerable following. This cooperation persisted for most of recorded history: the Roman persecution of the early Christians only lasted a few hundred years. Thereafter they cooperated. The social function of the religious organizations is developed in greater detail in Chapter B3 and will not be repeated here. Again, when a religious order gained power it tried to force everyone to obey its religious edicts. It was better to change religions than to be crucified or be burned. Most did that. The major deviation from this was the Jews who left Spain rather than become Christians. Another example is the spread of Arabs, which caused most people in their path to become Moslems. The Arabs made it very easy to become a Moslem. You simply had to utter a single phrase.

With the establishment of wealthy monarchs, nobles and the church, there was much support for architects, artists, musicians, writers, and creative individuals in many areas. This created openings for creative people and many adapted to these opportunities. As larger and improved ships were developed, explorers started to move around the world. People usually adapt to any new economic opportunities made available by technology. Those that do not adapt, remain moribund. Of course, being a country bounded by the seas or oceans is a big help.

15.3 Major Changes in Technology

Although the changes of food gathering and naval transportation were deemed to be technological, the major changes due to technology began with the Industrial Revolution. People left their cottage fabrication for industrial factories; wagoneers became locomotive drivers and sailing ship sailors had to learn how to work on steam ships. All of this illustrates the need to accommodate to changing technology. With the current Communications Revolution, learn how to use a computer, write codes, establish Web sites or become a minimum wage worker. As the evolutionists say, "adapt or die."

15.4 Applications of PL 8

15.4.1 Basic Rule for Adaptation

If you have lost the battle, it is better to accept being enslaved than becoming a dead warrior. Adapt to the existing environment or die and there will be no descendants. Whatever it is you represent will become extinct.

Another class of individuals finding themselves in a new environment, both physical and social, are immigrants, either legal or illegal. What follows is applicable to them too.

15.4.2 Learn the New Culture

Learn and practice the new language and customs. Learn all the modes of the new culture. If you want to retain some aspects of your original culture, do them in privacy and on your own time. Teach them to your children outside of whatever school system is enforced by the new society. If there is no applicable school system, teach your children the new culture as well, so that they can exist there successfully.

15.4.3 Learn the New Technology

You probably lost the battle because the conqueror had a more effective technology and/or organization than you did. Go to school and learn how to use that new technology. Maybe it's just how to aim and fire a rifle rather than an arrow. Or how to drive a motor vehicle rather than a cavalry horse. Whatever it was that made them win.

These days one has to learn how to use a computer—first how to enter data, then how to write programs and finally how to organize computer systems. Do not focus on a technology that is old and will soon disappear just because it is something you are familiar with and can do easily.

15.4.4 Basic Rules for the Future

No society is perfect and many of the customs and rules may be wrong. There was slavery for 5,000 years. Until recently, women were not allowed to vote. Today, homosexuality is believed by many to be a matter of choice and is sinful. The author believes that in the next generation homosexuality will be accepted as a genetic disorder. Thus, customs and laws change.

In all of the above, keep the eleven primordial rules, PLs, in mind and do not adopt any practice that violates them. They are believed to be the basis of a sound society.

PART C

Chapter 16 PL 9—The Safety of Diffusion

The scientific basis of this PL is that a specie had a greater chance of surviving natural disasters if specimens were spread out over large areas, even continents. Thus, a natural calamity such as an earthquake, volcanic eruption, or drought would not wipe out the entire specie. In addition, wide distribution places the specie into different environments, which by selection will increase the variability of the specie, which again is a good characteristic for the evolutionary process.

Homo sapiens rapidly spread around the world after its inception in East Africa. This diffusion occurred so long ago that there was enough time for local, physical environmental forces to select variation in the specie, i.e., skin color, eyelids, hair characteristics, genetic disease susceptibilities, height and facial differences sufficient to permit differentiation by visual recognition of the five different races of mankind. There is a counter-hypothesis that the different races represent separate passages at different times and places from the hominids to Homo sapiens. This theory was not well accepted at the time proposed (Coons 1962) because, in part, it implied a difference between the races that could not be explained by cultural influences. Today, the ability to measure DNA confirms the hypothesis of a single beginning.

The Earth being struck by a large meteor, which would change the Earth's climate for many years, is an event that may not be survivable by diffusion of the specie. But the specie is usually survivable by the ability of a small number of members to adapt to the new environment. This is why variability in the individuals of a species is so valuable if some of them may be able to adapt to and survive the calamity. While dinosaurs apparently were not able to survive such an event, many other species were.

However, many natural calamities are not world wide and diffusion of a specie over a large area is an insurance policy against extermination. This insurance policy even applies to societal calamities. Genocide cannot be 100% complete if the members of the unfortunate religion, clan or ethnic group have spread into many different countries. The Jews in America, and elsewhere, were an insurance policy against the success of Hitler's plans for the eradication of the Jews. Today, they are an insurance policy against the eradication of Israel by the Arabs. This is a societal force, rather than a biological urge.

16.1 Emigration

The desire to emigrate is reflected in Homo sapiens by the strong desire of many individuals to move from their family and friends, from the place they were born and grew up, to distant cities or even states and countries. (Lately this is more often a societal force as well as a biological urge.) Nature has selected the desire to roam as a human trait or characteristic because it serves as insurance against extermination of the specie.

Although the specie Homo sapiens, started in Africa, emigration of a few individuals or of a small group, dissatisfied for any reason could always go to the next valley simply because there was no one in that valley to object. They could only be prevented, and with great difficulty, by other members of their own tribe. It was rather easy to emigrate if one wanted to.

Today the ability to emigrate is severely limited by autocratic governments with modern technology; walls, barbed wire, troops and naval vessels. Witness the Soviet Union, Hitler's Germany, Cuba, Vietnam, etc. Today, the United Nation's Universal Declaration of Human Rights (Anon, 1948) includes the freedom to emigrate as a right of the individual and most of the Western World agrees with this. This basic right is covered in SciEthics by PL 5.

While evolution selected the desire to roam as a desirable trait for good reasons, this trait developed over several million years while the world was almost empty of people. Until about 10,000 years ago, at the time of the development of agriculture, the specie had immigrated to all the continents, except Antarctica, and had biologically adapted to their local environments. Those living in the tropics acquired more melanin in their skin as a protection against melanoma from the strong sun; those living in the northern regions lost their melanin so as to get the benefit of what little sunlight they had, and in between were people with varying skin colors and facial characteristics.

When, where, how and why other physiological differences between the five races of man developed is not yet clear. The main point is that the specie did migrate or diffuse all over the world. Whether we started from one point in Africa, or several points by coincidental similar mutations is not very germane. Since all the races can interbreed and produce fertile offspring, they are all of one specie.

16.2 Immigration

Now there is a dilemma. The next valley is not empty of humans and nowhere is the world empty of them. (Let's exclude Antarctica.) In fact, the world population is increasing extremely rapidly and the ability of the ecology to sustain the number of people has been questioned. Is Malthus finally correct? (Tanton 1996) And can society still maintain more people than it needs? These problems have led to many restrictions on immigration in many nations. Thus, we repeat the dilemma; does the right to emigrate include the right to immigrate?

Human rights were defined in Chapter B 7.1. Basically, the innate characteristics of humans as developed by evolution and reflected in our genes are desires, abilities or capabilities. These abilities do not guarantee life, liberty or happiness. Governments make and usually enforce a rule of human behavior, called a law. That law—but only if the government enforces it—ensures us a right to do a specific thing or to expect a specific benefit. Nature's Laws as applied to living things apply only in general to the entire population or to the average of large numbers and not to individuals therein. Thus, the word "rights," as applied to an individual, should be accepted only in the sense that a government has passed a law and will enforce it. Thus, although there is a genetic trait and desire to emigrate there is no right to immigrate into the next valley or promised land.

From an evolutionary scientific viewpoint, some immigration of members of the same specie is beneficial for that specie to survive locally for the following reasons:

a. Prevention of genetic deterioration due to inbreeding.
b. Increase in the variability of the gene pool.
c. Transfer of information that may increase local knowledge.

d. Accelerated growth of the size and strength of the receiving group, tribe or nation, i.e., America in the 19th century.

e. Creation of links with other groups and tribes.

f. Increase in the commonalty of language or at least the ability to interpret.

g. Finally, the joining of tribes into states and nations, although many other and stronger factors are involved in this step.

16.3 Social Diffusion

The next question is how many immigrants can a tribe absorb and still retain its culture? Single immigrants usually absorb the culture of the receiving group and become one of them. This is also a common practice in simian groups.

If a large number of immigrants are involved, they form enclaves, retain their own culture and in general it takes several generations for their offspring to be absorbed and for the original immigrants to die off. If immigrants of that specific ethnic group keep coming in, the enclaves or ghettos do not disappear.

If a very large number of immigrants are involved, they can swamp the indigenous group and replace them. This is especially true if the immigrants have a much stronger technology. The Americans, Australians and Japanese are relatively recent examples of these phenomena. Or they can interbreed and create a new culture. The Spanish in South America, Central America and Mexico illustrate this.

On what basis does one compare the U.S. today to the Indian tribes before 1492? Would they have developed the science and technology we have today in 500 years? Most likely not. Thus, from a societal evolutionary viewpoint, replacement of the American Indians, the Australian aborigines and the blacks in South Africa represent the growth of a stronger more productive society. The loss of Indians does represents a loss of diversity, which constitutes a potential danger. Have we left enough American Redmen alive to act as an insurance policy?

The above statements represent an attempt to generalize a very large and varied number of examples of the effects of immigration on a resident culture. Exceptions and other results can always be found. Immigration can yield benefits to the receiving country, especially if they diffuse into the culture. However, there is no "right" to immigrate. Restriction of immigration is an effort for self-preservation of the tribe. (Williamson, 1996) We are trying here to present a simplified scenario to use as a basis for establishing, on scientific evolutionary grounds, the lack of any "rights" of emigrants to immigrate.

If the number of immigrants is very large and they can totally wipe out the indigenous people, we may have lost some diversity, assuming the immigrants represent an ethnic group that already exists in many places. If the immigrants represent the last members of an ethnic group, then they represent a potential insurance policy. If the immigrants are single or a few individuals, then because of the advantages enumerated above, they should be admitted and absorbed.

If the immigrants are too many, then because of the disadvantages enumerated above, they should, if possible, not be admitted. Because of the potential advantages to the admitting country, admission should be based on the following characteristics of the immigrants:

a. Age and health, although deficiencies readily rectified are acceptable.
b. Education and/or intelligence.
c. Obstacles overcome in emigrating and coming to the receiving country. This would demonstrate the strength of the desire to come, perseverance and courage. Immigrants in small boats, who have suffered great hardships and the risk of death, are the kind of people a country should value and accept readily.
d. Willingness to learn the language and customs of the new country, yet maintain those of the country they left for cross linkage and personal enjoyment. This would be a measure of their intent to be absorbed rather than live ghetto style.
e. Citizenship and the right to vote should be tied to some measure of absorption and not just time in the new country. The melting pot model is better for the receiving county than the salad bowl model. The USSR and Yugoslavia/Bosnia are examples of salad bowl societies. Allowances should be made if the immigrant is too old to change. Cross marriages should be encouraged.
f. If the number of the group already in the receiving country is small, the immigrant should be more acceptable, if large, less acceptable.
g. Wealth and the commitment to establish industry.

The total number of immigrants allowed per year into a country should be a small fraction of the total population, say 1% or less, unless the country is highly under populated or losing population. The other consideration is the ability of a county to maintain its ecology and standard of living. Each country should study this question and determine an optimum current population and annual increase or decrease if currently over populated. (Note 3)

Nature always maintained a balance between the number of people in a given geography and the ability of those people to live and perpetuate themselves in that geography. If they were hunter-gatherers, the population density was limited by the flora and fauna in that geography. If they were agriculturists-herders, then a higher density of population was feasible. With the growth of science and technology, the maintainable population density became even higher. Whereas, formerly 75% of the population was required to raise food for 100% of the population, in the U.S. today only 5% do the same job. Thus, maintainable population density is a function of the technical advancement of the society.

A new limitation on total world population appears to be forming. This is the ability of the ecology to absorb the waste products of the society. In addition to the solid wastes associated with trash and garbage, this includes the tremendous generation of CO_2 caused by the use of oil and coal and the loss of forests that normally absorb CO_2. This problem can today be solved by technology or by population reduction. Ultimately there has to be population control. If we do not do so, Nature will. The Four Horsemen of the Apocalypse still ride, i.e., new viruses.

A strong factor against population self-control is that the tenants of religion, formulated when the world was relatively empty, was "Be fertile and multiply." Now, although the world is not so empty, religions are reluctant to change their maxims and adapt to the new conditions.

To summarize, there is no "right" to immigrate, although under some circumstances it is beneficial to the receiving country. The argument that all humans have the right to immigrate anywhere can be destructive to the culture, society and polity of the most advanced and attractive countries. There is an interesting novel of this danger. (Raspail, 1973)

Although there is no "right" to immigrate, during much of our past history, the militarily more advanced tribes and nations, larger and stronger by their more effective structure, usually moved into the next valley by force. This was usually justified by various self-serving slogans such as:

a. Might Makes Right
b. Manifest Destiny
c. We are bringing a higher civilization
d. Conversion to Christianity

In a Darwinian sense, this is a process of social evolution. Just as more effective human traits were selected biologically, so more effective social forms are selected by human force. This is more easily justifiable when there are obvious different biological characteristics. Some of these differences are visible skin coloration, eye shape and facial features. Others, like the susceptibility to specific diseases, intelligence and genome structure were not visible but can now be measured. However, until the advent of modern science, the very apparent differences were used to justify colonization worldwide.

Thus, in effect, the "Might Makes Right" process represents one of the selection processes in societal evolution. Examples are England conquering Southern Ireland and Wales to form Great Britain; the USA conquering the natives west of its initial formation or buying Florida, the Southwest and Alaska. Examples of ineffective conquering are the USA in the Philippines and Puerto Rico, Russia in Eastern Europe and Spain in South America. One of the major reasons for the ineffectiveness was the growing reluctance to destroy and replace the existing population, which is believed due to the growing realization that we are all one specie and the increasing moral standards.

16.4 Applications of PL 8

The applications are defined in the above sections and do not require duplication.

PART C

Chapter 17 PL 10—Private Property

17.1 The Price One Pays To Earn It

Long ago and far away, an individual pre-Homo sapien who took time off to shape a stone knife could have used that time to hunt for food or to chase females in heat. That was a large price to pay and the stone knife assumed the value of the neglected opportunities. Thus, the knife belongs to him. This is the birth of the genetic based emotional feeling for private property; I made it, it belongs to me. Very small children holding onto a toy exhibit this genetic tendency.

This has since expanded to items purchased with money earned in a job that has nothing to do with the item purchased. When one works to improve his/her owned house or farm, those efforts become part of their property. However, if the house is only rented, or the farm is part of a commune, then there is a great reluctance to put, time, energy and funds for supplies into improvements because that increased value won't belong to you.

The communist dogma of "from each what he can do and to each what he needs" completely violates this genetically innate feeling for private property. Why work hard to make six widgets a day if my coworker makes only three and we both get paid the same? This violates the innate concept of fairness. The Soviet Union philosophy and the way it was applied violate many of the PLs that assisted the evolution of Homo sapiens. It only took some 70 years for that social structure to fail. Social Darwinism is very effective; it is not dead.

17.2 But You Can't Take It With You

If the wealth or private property you have accumulated is small, then there is no problem. You leave it to your descendants. Small is defined as not enough to determine the rest of their lives. They still have to struggle for success.

If the wealth or private property you have accumulated is large, i.e., enough to determine the rest of their lives, then you do have a problem. There is a conflict between PL 3 and PL 10. From PL 3, one wants to ensure the health and longevity of their children and grandchildren regardless of their capabilities and situations. So, you want to leave them more than mere trinkets. This problem is covered next.

17.3 The Ease of Inheritance

One is strongly tempted to leave the wealth and property one has accumulated to one's descendants. To the extent that one's wealth is adequate, it should be used to ensure the health and education of one's descendants, i.e., give them the tools that are necessary for success. These tools are necessary, but not sufficient. In addition, they themselves must have the desire and drive to succeed. However, if you give them the wealth to be a multimillionaire when you pass on and they haven't started their post educational life, the desire and drive are gone. Such a gift really neutralizes the self-esteem realized when one's wealth is self-earned.

This problem is becoming better appreciated by wealthy people these days who set up foundations to promote their own concepts of social improvements and leave only sufficient funds to their heirs to ensure a comfortable life and the means to create their own wealth. This is a difficult decision because one always wants to give their descendants what ever they have accumulated.

17.4 Applications of PL 10

17.4.1 Appreciate, accept, and protect other people's right to their personal property.

17.4.2 Allow others to do whatever they want with their personal property as long as it does not create hurt nor harm to others now and in the future.

17.4.3 If you have acquired wealth and property, plan to leave just enough to your descendants to ensure their health and education. Do not leave great amounts so that their drive for self-achievement is destroyed.

17.4.4 With the excess funds, establish a perpetual foundation to further your own ideas for the betterment of society. This may also get you some immortality.

PART C

Chapter 18 PL 11—Public Property

18.1 The Physical Environment

For millions of years, Homo sapiens adjusted to changes in the physical environment over which he had little or no control. Gradually from about 10,000 to 15,000 years ago, mankind started to define his own environment. The domestication of animals, the development of agricultural and the utility of permanent habitats all began to define his physical and social environments. The most impressive social aspects were the development of cities rather than camps and complex governments rather than tribal chiefs. The effects of nature retreated to the weather, volcanoes, earthquakes and storms at sea.

18.2 The Physical Environment—Modified By Mankind

To a great extent, the physical environment today is almost totally defined by mankind. From ignorance of the cause and effect aspect of his actions on the physical environment, many results of his actions were and are still destructive. When the world was mostly empty of people, the things the few people did had no noticeable effect. This freedom to do anything that affected the physical environment became part of the social structure.

The physical environment changed by mankind, often destructively, was due to ignorance and the freedom to do anything with respect to the environment. Improper plowing led to erosion, over hunting and over fishing led to extermination of species, and the use of chemicals in agriculture and industry led to contamination of the water supply, rivers and lakes and now to the seaming endless oceans. Thus, mankind is destroying the physical environment in which he grew up to his present physical condition and which defined his genetic composition.

Most people in the developed countries live in cities, see only concrete walkways, streets and buildings, travel is by vehicles over concrete highways, railroad tracks, or in airplanes flying high and with no contact with the earth. Flying was not part of our evolutionary experience. We cross rivers on bridges rather than walk upstream until we can cross on a low spot. Only 5% of our population live on farms. The trend in all countries is for farmers to move to the cities. More and more of the Earth's population now live in an environment created by mankind.

In addition to damaging the flora of nature, man also damages the fauna. Over fishing has put many species on the endangered list. Cutting forests to get more farmland has destroyed many insect and animal species. Thus, mankind is destroying the physical environment in which he grew up to his present physical condition and which defined his genetic composition.

18.3 The Physical Environments—Not Affected By Mankind

While weather, volcanoes, earthquakes and storms at sea still occur without any control by mankind, the ability to predict them has increased to the point that their damage has become limited. Weather is mildly predictable, ships can sail around storms and populations can be advised to move away from impending volcanoes and hurricanes. The destructive effects of earthquakes can be minimized by applying structural engineering to buildings and transportation overpasses; preparing trained rescue teams and equipment; and by the rapid transportation of food, medical supplies and temporary housing. Earthquakes are still neither predictable nor controllable but with time science may be helpful even there.

Thus, to a great extent our environment is man made. Our knowledge as how to avoid damaging it and how to change it for the better is growing. Our ability to reduce the continuing damage and to avoid new damages is very limited because of population growth, economic forces, unawareness by the public, and on growing global economic dependence on exports to all the other nations. But both our ability to understand the consequences and to do something about it is increasing.

18.4 The Social Environment

Much of the damage to the physical environment was made possible by mankind operating in a social environment which placed a positive emphasis on economic success and did not, until recently, place any restraints on any activity that damaged the environments. In the same way, little thought was placed on the destructive effects of government actions on the social environment.

The social environment established by dictatorships violated many of the SciEthics and led to emigration and finally revolutions that destroyed the dictatorship. Whether it was the "Let them eat cake" of the French royalty or the "From each according to his ability and to each according to his needs" of the communists, the action of absolute rulers or dictators reflected an abysmal ignorance or rejection of the deeply rooted genetic tendencies of Homo sapiens. The responsibility of mankind for saving both the physical and social environments is included in greater detail later in Part D.

18.5 Applications of PL 11

We must recognize our responsibility to protect the physical environment. Some of the applications of PL 11 to the physical environment follow.

18.5.1 Obey the three Rs; reduce, recover and recycle all the material used at home, in business and in industry.

18.5.2 Encourage and support the development of non-polluting energy sources, i.e. solar and wind energy, electric and natural gas autos, busses and trucks. Encourage and support the development of rapid transport systems, subways, trolleys and railroads, i.e. reduce dependence on personal autos.

18.5.3 Encourage and support the exporting of food to tropical countries in order to avoid their necessity of cutting down rain forests; providing jobs by moving industry there. This is one aspect of the gradual globalization of the world's economy.

18.5.4 An obvious major corrective process would be population control. Raising the income level of all people everywhere to where the educational level reaches the point that the value of having only two children becomes self-apparent can painlessly attain this. As a birth level of 2.1 children per family is required in order to maintain a constant population, the population will eventually stabilize and then recede.

(Singer, 1999) It will be a tough road to this objective and it may represent an impossible goal.

18.5.5 There are many, many other things, less difficult and less idealistic that we can do to minimize our impact on the physical environment. The economics of each step generates the difficulty in taking each step. It is cheaper and easier to burn gasoline in autos than to create a gaseous hydrogen distribution infrastructure and a hydrogen-fueled auto.

PART D ENSURING A SUCCESSFUL FUTURE

Chapter 19 Biological Evolution of Mankind

19.1 Nature's Evolution Always Looks Backward

Evolution in the Darwinian sense reflects the selectivity of traits that are more effective in perpetuating the specie in the then prevailing environment. Some of these changes may create new specie. None of evolutionary changes are in anticipation of a possible future change in the environment. Thus, the PLs in Part C reflect human traits that were successful, in past environments, in perpetuating the specie and thus ensured that we would be here as we are.

Society is believed to also be acting, although unwittingly, in a biological selectivity mode. People who cannot behave in a manner acceptable to the rest of their neighbors tend to misbehave, commit crimes, become very poor and do not have families and children or if they do, are not very successful. Of course, many exploit the welfare system and that may be negative selectivity.

In addition, wars and colonizing exploitation tend to diminish and even eliminate long-lived clans and societies. The Australian Aborigines and the American Indians are still here, but in much smaller numbers. Again, this is looking backward at the 1500-1900 time period. Our ethical viewpoints have changed and we are no longer consciously trying to eliminate them.

19.2 Society As An Unwittingly Environmental Destructive Force

Society, until very recently, has unwittingly been acting as a force to change the physical environment as well as the social environment. There has been a total disregard to the extinction of species of plants and animals although there were enough examples of the benefits to be obtained from other species of plants and animals to warrant the preservation of all known species. In addition over farming (resulting from over population) has destroyed forests, extended deserts, eroded land, contaminated rivers and introduced dangerous chemicals into the oceans and inland water systems.

Over fishing has reduced many species, i.e., whales, to the point that fishing is almost no longer economical. Fishing farms may be a partial solution for preserving some species, but many are not suitable for this kind of life, i.e. whales again. Restrictions on whaling are becoming effective.

Society as a whole, which includes consumption as well as providing foods and goods, is generating huge amounts of trash and waste that are disposed of in dumps throughout the country. These dumps can become sources of dangerous gases. Yes, some dumps have been converted into power plants using the gas as fuel. There has been much environmental damage from manufacturing involving the discarding of byproducts and dangerous chemicals used in the manufacturing technology.

It is recognized there are many organizations that are aware of the above factors and agitate for their control. We do not want to minimize the beneficial changes their activity has produced. They have forced our government to establish the Environmental Protection Agency, EPA, and to enforce laws to avoid contamination. Many endangered species are coming back due to limitations imposed on economic activities. Most people now understand the dangers created by the modern economy, but are often unable to deal with them.

The question here however, is what is the effect of these physical environmental changes created by society on the evolution of mankind. Most of the effects are minuses. This author believes that Homo sapiens, or at least those at the upper end of the intelligence distribution, have the knowledge and the ability to anticipate possible future dangerous changes in the environment, and the means to reduce their impact and even prevent specie extinction. Is the remainder of humanity willing to accept and support advances in these precautionary activities?

19.3 Genetic Engineering On Humans For Medical Reasons

With the completion of the Genome Project, we will have the full blueprint of the genetic makeup of mankind. Today, even with only a partial knowledge, we are able to identify defective genes that cause illnesses and to provide fixes that offset their effect, if not change the genes themselves. However, there is no question that we will eventually be able to substitute good genes for defective ones.

Initially, genetic engineering will be used to prevent or cure diseases and deformities. This is just the extension of conventional medical practice and is widely accepted. Any cure for cancer would be acceptable. Prior to the development of a cure for any given ailment is the diagnosis of that ailment very early in the fetus. Then, if a repair is not available, the option of an abortion diminishes the continued spread of that ailment and allows the substitution of a more healthy baby for an impaired one.

This represents an acceleration of the evolutionary biological selectivity and is SciEthically acceptable. The arguments against abortion are based on religious concepts and are contradicted by the fact Nature causes many self-abortions in early pregnancy.

19.3 Genetic Engineering On Humans For Selectivity

With the information available from diagnosis of the fetus, it is now possible by abortion to select the sex of an offspring. Thus, in China, which tried to control over population by limiting couples to one child, the sex preferred by the culture was male and the use of abortion or infanticide of females resulted in increasing the male-female ratio to the extent that females became more valuable and the ratio started to settle down. With the now permitted two children per family, the problem is slowly correcting itself.

From a SciEthical viewpoint, there should be an equal number of males and females so that the freedom of choice, PL 5, exists. However, if a family already has two or more children of one sex, then selectivity of the sex of the next child is acceptable, also per PL 5.

With the ability to be selective about the sex of one's offspring, there is or will also be, the ability to select for beauty, hair color, i.e., for appearance. Is this ethical? It is applying the values of a societal culture on the biology of the offspring. To the extent that the desire for beauty and its value is deeply embedded in the psyche, it reflects a genetic basis and is acceptable.

Certain disfiguring disformities, such as club feet and split lips, can now be medically rectified. Would not avoiding the defect by genetic engineering on the fetus be more cost effective? If not correctable, then abortion would remove the problem.

If we have acquired the ability to be selective about appearances, would it not be preferable to select for intelligence, the pre-eminent characteristic that provided mankind with the ability to conquer the world? Attempts have been made to search the brain of Einstein and other great men to find similarities. Was there some blood saved so that now their genes could be used for the same purpose?

It appears that the slow pace of natural evolution, which takes hundreds of thousands of years to become effective, will eventually be replaced by the scientific ability to genetically engineer mankind. This is just the extension of man's abilities to change the environment that blindly modifies or eliminates the species to knowingly directly modifying the specie. While natural modifications of the specie are primarily changes that help to preserve a specie, or a modification thereof, it is without aim as regard to an ultimate goal.

The Law of Unexpected Consequences often produces bad features from good intents. Examples are the global warming from the extended use of hydrocarbon fuels for heating and transportation and the more rapid transmission of diseases by the ease of modern transportation, i.e., HIV and AIDS.

The real question with respect to genetic engineering is do we have the collective wisdom and ability to engineer changes that will ultimately produce Homo superiors and also ensure the perpetuation of the Genus. We, or rather our grandchildren, shall see.

PART D

Chapter 20 Societal Evolution

20.1 Borders

Borders between clans, tribes, city/states and finally nations usually depended on geography. Rivers, mountains, oceans and deserts often defined the boundaries between different social groups. With time, these boundaries became institutionalized and usually respected, although dominant societies often violated them, conquered the society across the river and incorporated them into their own society. England conquered the Welsh, Scots, and Irish and made them part of Great Britain.

With the advances in modern society, there was an increased flow of people and products between countries and the boundaries became less exclusive. Trade between countries has always been a strong factor in diminishing the isolation between different societies. Now advanced transportation has almost destroyed that isolation. What little remains is usually imposed by a dictatorial form of government, which will use any techniques to retain its authority.

Thus, the multifaceted aspect of boundaries is slowly diminishing and disappearing and we are approaching the One World envisioned by Weldon Wilkie in the first part of this century. The boundary between the U.S. and Canada, although formally maintained, is in fact almost nonexistent. The European Union has accomplished the same thing in Europe. The American culture, spread by TV and the Internet, is now widely visible worldwide although the French are still complaining.

This diffusion of economic practices is growing and will increase. It will also include social and cultural aspects such as religion, language, driving on the right side of the road and drinking Coca-Cola. Eventually, the United States will adopt the metric system. Thus, it is believed that although boundaries and differences may still exist formally and legally, they are in function and fact disappearing. We are slowly becoming one world.

20.2 Governments

The defeat and breakup of the Soviet Union, and the economic victories of the capitalist free market are reflected in the steady growth of democratic governments everywhere. This change of governments is nowhere complete and there are still autocracies with cronyism and corruption in many countries. However, the UN, the International Banks and the democratic countries are all applying pressure on such governments to economically open markets and thereby become democratic.

Manufacturing is shifting from the First World countries to the Third World countries because of lower labor costs. For this economic interplay to be practical, the Third World countries must operate under law and by the rules of the First World countries. Child labor is a no-no. Governments that are still autocratic are pretending to be democratic. It is believed that the greater efficiencies of the democratic countries will eventually make all governments copy their practices. It will take a long time for countries now ruled by religious fanatics, but it will eventually happen.

20.3 Conformity vs. Pluralism

In the growth of the United States, a nation of immigrants, the need for unity was emphasized by the melting pot concept in which immigrants would all learn English and follow the cultural practices established by the initial European immigrants. Their own prior ethnic practices could be followed in their own time with their own organizations, but not be supported with public funds. This idea prevailed through the first half of this century. With the subsequent inability to integrate the African-Americans, and to a lesser extent the Latino-Americans, the visibility of ethnic groups in public operations and political activity has grown and the melting pot concept was replaced by the salad bowl concept; where different ethnicities were mixed up but still recognizable. This diversity was justified as reflecting the fact that the United States was a nation of immigrants, and that democracy permitted, even encouraged this arrangement.

As long as each ethnic group followed PL 4, i.e., cooperated with their neighbors even if of another ethnicity and did them no harm, there is no objection to this as it reflects PL 5, the freedom of choice. From a functional view point it is obvious that all citizens of any state should be able to speak a common language, accept and obey the common laws and avoid the use of public funds for the benefit of specific ethnic groups. This is an extension of the separation of church and state concept to the separation of ethnicity and the state. The strength of a unified country is much greater than one composed of combative ethnic groups. (In this context, all citizens should serve a term in the armed forces to establish a commonalty and ensure efficient cooperation in times of emergency and warfare.) Immigrants who left their own country voluntarily to come to the U.S. should be prepared and willing to become Americans. It is recognized that elderly immigrants will find this difficult. They have been and should be allowed to form and live in ghettos, but their children should become Americans. Thus, from a SciEthical viewpoint public policy should be aimed at unifying the country rather than establishing separate ethnic groups that act as separate states. This is an interpretation and extension of PL 4.

20.4 Education

As part of the argument for unity in the population of any country, it is believed that the education of all the young members should have the following characteristics:

Be Uniform. All children should attend integrated public or private schools to learn a common language and a common culture. The separation of church and state concept should not allow religious schools acting as the sole school for children. If their parents want them to learn a specific religion or the practices of a specific ethnicity, then this should be done on their own time, i.e. evenings or weekends, and via their own organizations. Observation of religious or ethnic holidays, i.e., Christmas or St. Patrick's Day, should not be a legal holiday, although they have been accepted and are probably not feasible to change. Thanksgiving, the 4th of July and New Year's Day are acceptable legal holidays.

Be Available. Public paid for education should be required and available to all children through college. As capability to absorb an education is not the same for all, and probably distributed in a curve of some sort, then schools should be diverse enough to accommodate all students and with different goals. Whether this should be done in the same facility or in different ones is a difficult problem. However, the ability to educate some better or more than others should not be abandoned in order to provide a common educational environment.

<u>Be Competitive</u>. To provide parents with the freedom of choice, PL 5, all children should get vouchers to attend any school their parents select, excluding religious schools. If a religious organization wants to provide an integrated school without religious indoctrination, that should be acceptable. Competition between schools would lead to more effective schools and monopolies would become expensive and ineffective as there is no mechanism of negative feedback.

<u>Emphasize Intelligence and Creativity</u>. The great driving force for the advancement of Homo sapiens has been the growth of multifaceted intelligence or creativity, per PL 6. Thus, it would seem that our educational system should emphasize these aspects rather than others. Children displaying creativity should be encouraged and provided scholarships to ensure they will continue with their education regardless of the economic conditions at home. Most parents would be glad to see their child incentivized and even sent to a live-in school in order to ensure that the child does better than they have done. Let us emphasize intelligence and creativity instead of the strength, facility and speed favored by sports.

PART D

Chapter 21 Societal Responsibilities

21.1 Ecology Control

The freedom for people to do whatever they wanted with regard to the physical environment was an exercise of PL 5 as long as the results had no effect on their neighbors, PL 4. The fact that this was no longer true was brought out strongly by Rachel Carson in 1962 and the arguments for protection of the environment have grown stronger ever since. There is now no question that the environment of the world is being badly damaged by the consumption of hydrocarbons and the disposal of wastes by industry and the consumer. This is due both to the population growth and to the increase in consumption resulting from advancements in technology and the increasing wealth of the population.

It is now the responsibility of society to control this problem. Individuals, each acting in their own self interest, cannot be expected to do this. The three R's are applicable; Reduce, Reuse and Recover. In addition, technology should be directed at solving this problem. Examples are increasing the effectiveness of harmless waste disposal, substituting solar energy for hydrocarbons and eventually dumping harmful wastes into the sun. But the problem is largely governmental; providing convenient rapid transit to reduce the use of automobiles, establishing waste-sorting systems in order to increase the recovery of useful materials as well as to provide low-skill jobs for those on relief.

How to enforce such steps in Third World countries as well as in First World countries is a difficult problem, but it must be solved. Using Third World countries as dumps is not acceptable, as it violates PL 4. However, damaging the ecology of each country and of the world must be slowed and stopped. It is the responsibility of mankind to do this in order to avoid the potential danger of extinction of Homo sapiens or the destruction of our advanced society.

21.2 Population Control

The ecological problems arising from the growth in the world's population has been covered several times. There are many potential solutions. Passing laws to reduce the number of children a couple has been tried (Note 3), but is difficult to enforce as it violates PL 3. Increasing the wealth of a couple is more effective as the number of births per couple in advanced countries has dropped to two or less. This approach is happening as we are slowly becoming One World, but it will take lots of time and will cause a shift in the ethnic composition of the world's population.

Many other steps can be taken so that people voluntarily reduce their own number of children. Some of these are:

a. Provide sex education.
b. Make contraceptives readily available.
c. Make voluntary sterilization readily available.
d. Make safe abortions readily available.
e. Provide marriage status for homosexual couples.

f. Remove tax advantages for having many children.

g. Provide a social security system for older people so that they do not need to have many children who can support them later.

h. Let couples select the sex of a child so that they can have one of each kind and do not need to keep trying to do so.

It is recognized that many of the above items may have a small (but not negligent) influence on the population control problem. However, the desire to have children is so deeply rooted in our genes that only voluntary steps are practical. However, we are having a population explosion problem and society must do all it can to control it.

21.3 Alternatives to War

In an objectionable way, war has been an effective societal selective process. As one of the Four Horsemen of the Apocalypse, it has also acted to reduce the world's population. Without a One-World government and an international court system operating under an international set of laws, wars have been the only way to resolve conflicts.

The American Civil War in 1860 is a good example. No one objects to good wars, i.e., the American Revolutionary War with England, WWI and WW 11. Of course, we now question the Vietnam War, the war with Spain and with Mexico during the American expansion, and the wars with the native American Indians. But no one wants to return the southwest states to Mexico or Manhattan Island to the Indians.

The concept of a nation's sovereignty implies the ability of one nation to say no to another nation. Thus, until there is an effective world government, there is no way to avoid wars if negotiations are ineffective. This is another example of the practical rule, "Might Makes Right." Olympic sports have been interpreted as a substitute for war, but there is no known situation in which a dispute between two nations has been so resolved.

In view of the drastic differences between the governments and societal systems of the many nations now existing, it is difficult to see the United States accepting control of its activities or that of its citizens abroad based on a majority vote of all the members of the UN. It will be a long time before there is a legally effective One World system. It will be preceded by a growing informal, economic and cultural uniformity.

21.4 Prevention of Extinction of Homo sapiens

The primary duty of all people and all current governments is to prevent the extinction of the species, Homo sapiens. There are many things we can do. Some of them are given below.

a. Increase and extend a world-wide program to detect viruses. As part of this program, do research on vaccines for variations of viral microbes and viruses that show up periodically. Use existing vaccines world wide, to everybody as a global expense. If we can eliminate viruses world wide, we will diminish the possibility of variants developing that are immune to our vaccines. Also do research on how to increase the ability of the human immune system to counteract any kind of virus.

b. Take the danger of climate warming as a result of the CO_2 increase seriously. Recognize that the Third World countries cannot afford the steps that the First World countries can take and subsidize them. World wide, replace power plants that use hydrocarbon fuels with nuclear power plants. Solve the nuclear waste disposal problem by using cargo missiles to fire the waste directly into the Sun.

c. Establish a continuous and thorough astronomical search for meteors that might strike Earth. As part of this program, develop a series of rocket-powered vehicles with atomic warheads that can be used to deflect or destroy such a meteor. As another part of this program, practice using these vehicles on a meteor that will closely pass but not hit Earth. This program could constitute a use for all the atomic warhead and ballistic missiles now wasting away.

d. Extend the diffusion of Home sapiens to large self-sustaining space stations and eventually to permanent colonies on Mars and Venus. In preparation for the latter two, do research on terra-forming both planets.

It is recognized that the above suggestions will be very difficult to accomplish or even start. Some of them even look like science fiction. (But science fiction has often foretold what can and will happen.) They are presented as examples of what we, as modern Homo sapiens, should be doing to preserve our specie. There are probably additional applicable efforts.

PART D

Chapter 22 Conclusions

The immediate goal of our ethical system is to ensure the continued existence of our specie, Homo sapiens. We have the broad knowledge to do this. There is even a secondary goal; to produce an advanced specie, Homo superiors. While the details are not complete, there is an awareness of the broad path we must take to achieve both these goals. The basic problem is that our culture has not accepted these goals, especially the latter one, and significant resources are not being used to achieve them. Again, the knowledge and the goals exist mainly in the few percent of the population at the high end of the intelligence distribution. Although secularization is increasing, the bulk of the population still believes in gods, souls in ova just fertilized, and having as many children as possible. But we now know about the DNA chain and its relation to human behavior. So the knowledge is there and it will spread, albeit slowly.

There are many obstacles to advancing the program to prevent the extinction of Homo sapiens. Most people think it is a very, very unlikely event, although the dinosaurs are a good example. The continuing loss of species mainly bothers the scientists in the biological fields. The general public is not impacted. The inertia of the culture to events of very low probability or the loss of an obscure specie is a major obstacle. There is always the concern that the government is wasting our money on useless projects.

Thus, we have a long way to go to make any of these steps a meaningful program for these two goals. But there have been starts. It generally takes a catastrophe before the general public accepts a new responsibility. Although we have started on many of these programs, there is still a long way to go. But we have started, and we should and can finish them.

PART E EPILOGUE

Chapter 23 Postscript

23.1 Not Cast In Stone

Like all scientific hypotheses, the SciEthical approach is not cast in stone because it was delivered by a supernatural entity. Neither is it complete. Many issues have been omitted in order to stay simple and clear. It is subject to modification, clarification, deletions and expansions. But these changes must be based on new or overlooked data. It is acknowledged that not all statements have been supported by a reference. But unlike philosophical concepts, it would not be changed because of "logical" arguments.

23.2 Suggestions Solicited

The reader is encouraged to submit suggestions, additions, references, additional examples, questions, counter-interpretations and comments. Different opinions are not wanted unless they are supported by objective data. If the material submitted is used, you will be given credit. Material should be sent to the distributor who will forward them to the author. An attempt will be made to respond to each of these communications.

23.3 Second Edition

It is believed by the author that the SciEthical approach presented in this book will serve as a basis for a secular ethical system for the next century and (millennium) and will be used to determine specific ethics for many situations not covered. Thus, it is planned to incorporate all useful submittals into a second edition.

BASIC ASSUMPTION & AXIOM

BASIC ASSUMPTION: The most precious thing in our human life is life itself.

<u>AXIOM</u>

Those physical and mental traits common to most humans, preferentially selected by natural and societal forces in the evolutionary process are positive guides* to acceptable actions, because they gave us our most precious human possession, life itself.

*These positive guides are considered to be the scientific basis of ethics.

Figure 1

111

SUMMARY OF SCIETHICS

1. Do what is necessary to stay alive.

2. Consenting safe sexual activity is healthy, but

3. All progeny must be nurtured until they are able to be self-supporting.

4. Cooperate with and do no harm to others.

5. Do your own thing, but remember No. 4 above.

6. Value intelligence* above strength and speed.

7. Speak up and give others the benefit of your thoughts.

8. Adapt to changes in the physical and societal environments.

9. You are free to leave your valley.

10. The things you make by your own efforts are yours.

11. Public property is for everyone's benefit and must not be damaged.

* Includes creativity in all fields.

Figure 2

THE TEN COMMANDMENTS

1. I am the Lord, thy God

2. Thou shall have no other god before me

3. Thou shall not take the name of the Lord in vain

4. Remember the sabbath and keep it holy

5. Honor thy father and thy mother

6. Thou shall not murder

7. Thou shall not commit adultery

8. Thou shall not steal

9. Thou shall not bear false witness

10. Thou shall not covet thy neighbor's possessions

Figure 3

GLOSSARY AND DEFINITIONS

1. Words are primarily labels for something else. The something else may be:

Objects. Things that can be seen, smelled, touched, heard or tasted; that are detectable to our five senses are called objects. Technology may extend the range of a sense so that we can, in effect, see things that we normally cannot see. X-rays and MRIs represent extensions of our ability to see. Audio amplifiers represent an extension of our ability to hear low sounds and very high frequencies. There are technical amplifiers for all our senses.

Physical Processes. When objects physically interact with and affect one another and/or form a new object, it is called a physical process. Sometimes a specific object changes, such as when a person ages, without any apparent interaction with other objects. However, the environment and internal biological objects called genes usually cause these changes. Thus, ice melting by itself or a person aging are also processes.

Mental Processes. Ideas, emotions and desires that are the result of complicated physical processes between different parts of the human mind are called a mental process. This could be either in the conscious or the subconscious mind. The results of this process in the subconscious mind can be kept in secrecy from the conscious mind.

Behavior. Physical acts by humans who reveal to others ideas, emotions or desires by movements of the body or parts thereof in order to effect a desired result are called behavior. This includes speech, body English and physical force.

2. Words are extremely efficient tools for us to communicate with each other. Try describing an eclipse without using the words "sun" and "moon." Words are labels for the objects, processes, ideas and feelings to which they refer. However, words have acquired characteristics of their own in addition to being merely labels. Just read on.

3. **The Evolution of Words**. With the growth of language, new meanings are ascribed to old words and new words are invented outright. Lazy authors use familiar words in a new sense, which is a hazard to the reader who thinks of the old meaning. More innovative authors compound new words from parts of old ones, from dead classical languages or even from obscure current languages. While this leads to less confusion in one sense it has a danger for the reader with a short memory. Although a glossary is often provided, too many new words create a text that reads like a foreign language. This is a common problem for someone reading a highly technical or legal text filled with many unfamiliar words.

Thus, for scientific or pedagogical clarity, key words should always be defined in the sense that the author intends to use, if that differs from the standard dictionary definition or from the prevalent common usage. This is of extreme importance when presenting a new theory or concept. Really new words are often required.

4. Pejorative Words. With usage, some words become endowed with characteristics other than those of simple labels. Motherhood, love, apple pie, God and Country, dirt, kindness, compassion, fairness, alien, vile, shitty, etc. are all able to arouse the emotion associated with the referent even if the referent is not even involved. The label has become a substitute for the emotion associated with the referent and signals one's attitude, opinion, judgment and final decision even before the facts can be presented, considered and judged. These words, called pejorative, essentially indicate a pre-judgment opinion and their use should arouse one's suspicion.

In order to avoid the emotional reaction to the use of these words, scientists instead use cool and detached words that are not pejorative, but act as simple descriptive labels. Politicians campaigning for office do just the opposite. Preachers from the pulpit do likewise.

5. Judgmental Words. There is much confusion about the words "good," "bad," "wrong," "right," and others that are judgmental in nature. For the reasons given above, these words are defined as follows:

> **Good/Bad/Neutral**. The judgment that an object or process promotes/ hinders/does not affect mankind's ability to adapt to the physical and/or societal environment.
>
> **Right/Wrong/Neutral**. The judgment that a human's behavior enhances/diminishes/ does not affect mankind's ability to understand reality, i.e., the physical and/or societal environment.
>
> **True/False and Correct/Incorrect**. The judgment on a statement that can be proved/disproved by scientific methods.
>
> **Uncertain/Nonsense**. Judgment on a statement that cannot be proven at this time by scientific methods because the tools are not available, i.e., telepathy/a perpetual motion machine will never be available because it is in absolute conflict with the precepts of science.
>
> **Moral/Immoral/Amoral**. Judgment on an action by a human that is Approved/Disapproved/Without moral content based on religious tenets given by a supernatural being and as interpreted by a human prophet or scholar.
>
> **Ethical/Unethical/Neutral**. Judgment on an action by a human that is Approved/Disapproved/Without ethical content based on philosophical tenets established by a human scholar.
>
> **SciEthical/Un-SciEthical/Neutral**. Judgment on an action by a human that Helps/Hinders/Does not affect a person's ability to follow the Primordial Laws (PLs) of SciEthics.

6. Definitions Used In This Book

> a. Moral: A rule for human behavior based on a specific religion. Other religions may use the same moral or a different one to define the acceptable human response to the same situation.

b. Ethic: A rule for human behavior in a given culture about a given situation usually based on a philosophy or custom.
c. SciEthic: An ethic based on science, as developed herein. There may be duplications and differences with a. and b. above.
d. Mankind: This includes men, women, children, the five races and any member of the specie "Homo sapiens." Technically, if the mating of a male and female results in a fertile offspring, the male and female are members of the same specie.
e. Hurt is used to identify physical damage to the person.
f. Harm is used to identify damage to objects and/or finances.
g. Offend is used to identify a violation of concepts, morals, ethics and feelings. It does not include physical or fiscal damage to the individual.
h. The basic family is an adult male and female and their children. There are many variations on this basic structure of a family.

As examples:

One or more children may be adopted.
There may be a few females as in polygamy.
There may be a few males as in polyandry.
The family may be extended to include grandparents and other close relatives.

In general, the male protects and provides the food and impregnates all the adult women who are not related to him. The female births and takes care of the children, prepares the food and maintains the household. Again there are many individual variations, but these do not invalidate the generalizations that describe the average family.

NOTE 1 DICTIONARY DEFINITIONS
Webster's New International Dictionary of the English Language 2nd Edition Unabridged 1951

Ethic 1: Ethics, also an ethical system. Ethic 2: Character or the ideals of character.
Ethics 1: A treatise on morals, specifically, The Ethical Works of Aristotle, three treatises: Nicoomachean Ethics; Eudemean Ethics; Magna Moralia. Ethics 2: The science of moral duty, the science of the ideal human character and the ideal ends of human action.

Webster's New International Dictionary of the English Language, 2nd Edition Unabridged 1951

Moral 1: Characterized by excellence in what pertains to practice or conduct; springing from, or pertaining to, man's natural sense or reasoned judgment of what is right and proper.
Synonyms: Moral, ethical
Moral may refer to either the science or the practice of right conduct. **Ethical** commonly suggests the science only. As applied to persons, **Moral** connotes right practice.
Ethics: The assent to right principals; an ethical system. Examples: an ethical system, a moral act.

<u>Webster's Ninth New Collegiate Dictionary 1985 pg. 426</u>

Ethic 1: The discipline dealing with what is good and bad and with moral duty and obligations.

2a: A set of moral duties or values.

2b: A theory or system of moral values.

2c: The principles of conduct governing an individual or a group.

<u>Webster's Ninth New Collegiate Dictionary 1985 pg. 771</u>

Moral 1a: of or relating to principles of right or wrong in behavior.

1b: expressing or teaching a conception of right behavior.

1c: conforming to a standard of right behavior.

Synonyms: Moral, ethical

Moral implies conformity to established sanctified codes or accepted notions of right and wrong.
Ethical may suggest the involvement of more difficult or subtle questions of rightness, fairness, or equity.

The above definitions assume that the knowledge of what is right and wrong exists. There is the hidden circular argument that acting morally is doing what is right and that doing right is doing what is moral. Note that the word "religion" is not involved. Moral and ethical are considered synonymous.

If one disregards the fact that the dictionaries consider the word moral and ethical to be synonymous, it seems to this author that the dictionary difference between ethics and morals is that ethics refers to a system and the science that generates the system, while morals are the acts of doing what that system specifies. Ethics are the general theory of "preferred" human behavior, and morals are the specific acts that humans should do. The definition of right and wrong is whatever the current culture and/or custom says it is. The definitions used in this book do not agree with the above definitions, however, they are more in line with the current usage in our culture and are clearly separated.

NOTE 2 GLOBALIZATION

The growth in size of political nations seems to have halted or even reversed after WWII with the demise of the Soviet Union and the breakup of Tito's Balkan Empire. There is no apparent "Roman Empire" in the making.

Instead, an economic path is apparently being followed in the evolutionary growth of governmental entities. Exports of American products and services are everywhere and local cultures are slowly becoming Americanized by them and by USA television programs. This is a peaceful "invasion" resisted mainly by the French who fight desperately to maintain their language and customs.

In addition to this growing foreign trade between all countries including Asia, there is a feeling of responsibility for the currencies and stock markets of all nations. The IMF and USA bailed out Mexico in the early '90s and bailed out Thailand, Indonesia and South Korea in the late '90s. Japan may be next. As condition of this bailout, the open market practices of the west and the abandonment of nepotism are being imposed. (Marshall, 1998).

This peaceful economic path replaces the martial occupational path of Imperialism, but the result is the same. One culture is being imposed on the world. This is the natural evolutionary growth of tribes to city-states to nations and finally to the One World of Weldon Wilkie. The detailed structure of this One World cannot be foretold, but it is not really critical. From a SciEthical viewpoint, the One World is inevitable and we should work toward a practical, efficient and long-lasting version.

The movement toward economic globalization has broadened to include non-economic topics. There have been global meetings, such as the Kyoto Conference on the responsibilities for ecological dangers; international meetings on sea pollution; wildlife preservation; whale hunting limitations and other non-economic topics.

The UN is beginning to reflect an attempt to establish a global law and a world court. Many private organizations (NGOs) as well as the UN try to alleviate poverty. Many of these efforts are only partially effective, but they all represent steps toward a transfer of wealth and practices proven effective, from the industrial nations to the poorer nations with ineffective practices. The longest journey starts with a few small steps.

NOTE 3 CHINA'S POLICY ON POPULATION CONTROL

While almost all governments acknowledge the problems associated with over population, few have made any serious attempt to control it. In democracies, independent groups advocate birth control and limited family size. However, religious groups are usually opposed to birth control—the Catholic hierarchy particularly. In most countries, the decision on family size is left to the individual. The absence of social safety nets usually leads to having many children for their support in old age.

China is the most heavily populated country with over one billion inhabitants. This autocratic communist system actually tried to do something about its overpopulation. It established a firm policy to limit family size. Any couple is limited to one child. Initially, this was strictly enforced by the law and structure of a totalitarian government. Contraceptives were widely distributed and abortion utilized to maintain the limit of one child per couple.

Recently, enforcement has been weakened and many violations are occurring. An unexpected by-product has been the preferential abortion of female fetuses as sons are preferred. This is causing a sharp decrease in the female to male ratio.

The enforcement of the one child per couple rule violates the evolutionary drive to have as many children as possible. Thus, the totalitarian attempt to curb population growth is not the right way. Over population is a very complex problem. Widespread wealth plus social safety nets may be the best answer.

NOTE 4 EVOLUTION ALTERNATIVES

There are two alternate theories on how the races of mankind evolved. In the first theory, there is no question that they originated in northern Africa and spread throughout the world. They are all members of the same specie, Homo sapiens, and can produce fertile offspring regardless of the different races of the parents. In this theory, they then changed to adapt to the local environment. In support of this theory is the example of dark skin for Africans to minimize the effect of strong sunlight and light skin for Scandinavians for maximize the effect of weak sunlight.

In the second theory, each race of mankind represent a separate crossing of the pre-hominids to Homo sapiens. For more detail on this subject see (Coons, 1962). The evidence from subsequent DNA does not support this theory.

From the viewpoint of SciEthics, it is immaterial which theory is correct. The behavior of all races exhibits the same natural evolutionary foundations but have different societal evolutionary paths. The differences in the visual outward appearances have been used to define the five races, but there are internal differences as well.

In the evolutionary context the more variability we have, the higher are the odds that one of the races will survive a natural calamity, thus ensuring the continuation of the specie.

NOTE 5 ADAPTATION TO A STRESSFUL SOCIAL ENVIRONMENT

An interesting successful example is the adaptation of Jews to a stressful societal environment identified as anti-Semitism. For over 2,000 years, Jews have maintained a tribal association and a social system under extremely onerous conditions and without a home country. The reasons for this success is that in addition to observing the first four essential PLs as defined by SciEthics, they also followed several of the following PLs.

PL 5 - Pursuit of of Happiness. Their ability to do this was largely greatly hindered by the rules imposed on them by anti-Semitism.

PL 6 - Intelligence and Curiosity. This was highly emphasized by their culture in the study of the Talmud, education of the males and social esteem and recognition for highly intelligent Rabbis. This apparently was a factor in that Jews generally have a higher IQ than other Europeans. (Herrnstein & Murray, 1994)

PL 7 - Free Speech. This was encouraged and practiced, but restricted to their own tribe.

PL 8 - Adaptation to Change. This was always done, no matter how onerous the change if they could not emigrate. Better to live than throw in the sponge.

PL 9 - Freedom to Leave. Although sometimes highly restricted, they did emigrate throughout the world. This served as a safety against extermination from local Russian pogroms and the Europe-wide Nazi genocide.

PL 10—Private Property. This was fully followed to the extent permitted. However, when a chance to emigrate occurred, they had no reluctance to leave property behind.

PL 11—Public Property. They had none during the 2,000 years of wandering around the world without a country.

This is the sole example of a tribe staying alive and united for 2,000 years without a country. A more complete and elegant presentation of this example is contained in the book, "The Jewish Mystique." (Ven de Hage, 1969)

NOTE 6 EXCEPTIONS

There are always exceptions to any general rule on human behavior. Thus, it may be useful for a man to have sex after forty if he has lost his family or not had any and now marries a younger woman. With modern medical science, a woman can still have a child after menopause if she has had some of her ova refrigerated and has them impregnated in a glass dish and then implanted in her womb.

There are many variants on the theme that exceptions do not invalidate a general rule applied to humans. It is only in the physical sciences and mathematics that an exception must be explained or the general rule must be modified to account for it.

NOTE 7 WARFARE

Wars were deliberately left out of this discourse because we seemingly have always had wars, which do constitute a way to resolve differences in a lawless world. Homo sapiens have evolved socially by the use of force. The Romans created a Pax Roman, which lasted a long time but failed to resolve the social and political problems arising from corruption and malfeasance. The invaders from outer space (Eastern Europe and Asia then) cleared the field, which allowed alternate social structures to evolve. Would we have had the Renaissance, the Reformation and the Enlightenment if the Pax Roman had persisted?

NOTE 8 HURT, HARM OR OFFEND

Hurt means physical damage to a person. Harm means damage to someone's property or assets. Neither hurt nor harm apply to actions that offend another person's ideas, feelings, beliefs, thoughts or concepts.

If one expresses a viewpoint that goes counter to another person's religion, that person may feel offended, but he/she has not suffered any physical hurt or financial harm. To avoid doing so would prevent the introduction of any new religion.

Thus, while SciEthics forbids hurting or harming your neighbor, it does not restrict your right to free speech as is covered in Chapter C 14, PL 7.

NOTE 9 CHRISTIAN ETHICS

An attempt was made to develop a list of Christian ethics by reviewing the book, "History of Christian Ethics-Vol. I From the New Testament to Augustine" (George W. Forell, 1979).

The single basic ethic from the start of Christianity was, "You Shall Love Your Neighbor As Yourself" from the Sermon On The Mount by Jesus Christ as reported, differently, by his various disciples. This order to love your neighbor was both a reflection that God loved all his people and a proof that you loved God.

The early Christians, who in response to this commandment, asked the question, "What Must I Do" needed specific actions to comply. The specific actions regarding one's love of their neighbor were initially a reflection of the ethics of the Hebrew Bible, but with a disregard for many specific laws that were hard to follow (Circumcision, food restrictions, the Sabbath observations, etc.) and did not reflect local customs and habits. Thus, the ethics and common sense of the local environments were also incorporated as this made conversion easier.

Eventually, the officers of the religious organization, philosophers and authors further developed the ethics in the light of the Love of God. It became very philosophical, complex and almost completely based on reasoning although common sense was a factor.

The reference did not contain a comprehensive list of ethics but there were lists of words labeled good and other lists labeled bad. The same deed could be listed as good and bad depending on the intent of the doer. The rational of the text became very convoluted and came back to the doer's relationship to God. Some of these words are listed below.

Good Bad
Love is the central norm Immorality
Joy and peace Impurity
Patience Licentiousness
Kindness Idolatry
Goodness Sorcery
Faithfulness Enmity and Strife
Gentleness Carousing
Self-control Jealousy
Admonish the Idle Anger
Encourage the faint hearted Wrath
Help the weak Malice
Women can be fellow workers Slander
Physical work Foul talk
Do not kill Do not lie
Do not commit adultery Stealing
Do not bear false witness Bearing false witness
Honor thy father and mother Defrauding
Leave vengeance to the Lord Charity is the end product of love

This author did not attempt to do a careful comparison of the definitions of all these words or short sentences. The amount of material in the text is enormous and the reasoning is philosophical and not scientific. The emphasis is mainly on the history of the development of ethics and not on the explanation of what the ethics mean in any detail. No attempt was made to read beyond Augustine, the end of the Volume I. Volumes II and III were not reviewed.

It seems that the basic Christian ethic, "Love Thy Neighbor as Thyself" is an appealing and significant alternative of PL 4, "Cooperate with your Neighbor and Do Him No Harm." The basis for this Christian ethic is that since God loves all his creatures, you should love them too or you would contradict your God.

With the development of the monotheistic GOD with separate prophets, Abraham and Moses for the Hebrews, Jesus for the Christians and Mohammed for the Muslims, the explanation shifted some to understand how the common man could get the benefits from believing in the tenets of their religion.

The SciEthic approach is that as man's intelligence grew, it included curiosity about reality and creativity trying to explain it. Thus, gods were invented to explain Nature as the acts of gods, and since gods could do anything, the solutions to real problems could be solved by gifts to the gods and asking for their help.

However, SciEthics accepts religions as very powerful productive activities that by trial and error developed ethics that in large part overlapped some of SciEthics. It was often excellent, but many times incorrect and over aggressive—killing Muslims during the Crusades, Jews during the Inquisition, pogroms, and the Holocaust. Christians could not accept that other worshippers of their One God would reject the concept that Jesus was the Son of God, as such a belief contradicted the basis of Christianity.

NOTE 10 THE POPULARITY OF SPORTS—(Suggested by Stuart M. Gordon)
The popularity of sports is due to the genetic selectivity during the hunter/warrior stage of evolution when speed and strength were important. The practices of young males emphasized speed, strength and dexterity in non-lethal competition with other males in their tribe when not engaged in hunting or warfare. These activities later became identified as sports. When men grew older and lost those physical abilities, they watched and judged the younger males who were still physically competitive. Thus, the interest in sports reflects another "over kill" of a genetic characteristic that was significant more than 10,000 years ago. (Do the international Olympics Games reflect a cultural extension of this trait to replace warfare? RG)

REFERENCES

Chapter Name and Year Title Publisher

7.1.2 Allman, John, 1998 Evolving Brains Scientific American Library

16.1 Anon, 1948 Universal Declaration of Human Rights UN, New York

10.2 Benedict, Ruth, 1934 Patterns of Culture Penguin Books

21.1 Carson, Rachel, 1962 Silent Spring Houghton Mifflin Co., Boston

2.5 Cattell, Raymond B., 1987 Beyondism: Religion From Science Prager Publishing

16.0 Coons, Carleton S., 1962 The Origin of Races Alfred A. Knopf, New York

2.6 Dugatkin, Lee, 1999 Cheating Monkeys and Citizen Bees The Free Press

4.4.1 Erler, Edward J., 1984 Equality, Natural Rights and the Rule of Law: The Claremont Institute, CA

Note 9 Forell, George W., 1979 History of Christian Ethics, Vol. 1 Augsburg Publishing House, Minn.

11.4.1 Frankel, Charles, 1951 The Social Contract by Rousseau Hafner Publ. Co., NY

12.4 Herrnstein & Murray, 1994 The Bell Curve The Free Press

4.2 Hobbes, Thomas, 1651 The Liviathan Oxford Press, 1955

4.2 Locke, John, 1690 Two Treatises on Government; Encyclopedia Britanica, 1952

Note 2 Marshall, 1998

16.3 Raspail, Jean, 1973 The Camp Of The Saints Social Contract Press

10.4.5 Sanger, Margaret, 1925 Happiness in Marriage Cromwell Press, 1926

18.5 Singer, Max, 1999 The Population Surprise The Atlantic Monthly, Aug. '99 pgs. 22-25

16.2 Tanton, John, 1996 Immigration & The Social Contract Aldershot, England

4.1 Tarnas, Richard, 1991 The Passion of the Western Mind Ballantine Books, New York

Note 5 VanDenHage, Ernest, 1969 The Jewish Mystique Stein & Day, NewYork

1.5 Visotzky, Burton, 1996 The Genesis of Ethics Crown Publs. Inc. NY

16.3 Williamson, Chilton, 1996 The Immigration Mystique: America's False Conscience Basic Books, New York

2.6 Wilson, Edward O., 1975 Sociobiology Harvard University Press

10.1 Wilson, Glen D., 1982 The Coolidge Effect Morrow Publishing

ABOUT THE AUTHOR

The author was born on June 23, 1917, in Brooklyn, New York and raised in the lower East Side of Manhattan and later in the west Bronx.

He studied engineering at the Cooper Union in New York City. He graduated in 1940 with a BS in mechanical engineering. His first exposure to rocketry was via a senior study project on rocket powered airplanes. Gordon concluded that they were not feasible. Exposure to V-1s and V-2s in Europe during W.W.II and participating in a raid on Penemunda showed that he did not deserve an A for his senior project.

He joined Aerojet Engineering Corporation in Azusa in July of 1945. He worked on liquid propellant rocket engines and rotating space power systems. He refused an assignment to a nuclear project because he knew nothing nuclear and was sent back to school in July of 1955.

At UC Berkeley, he learned all about neutrons, protons, electrons, and nuclear reactors. Gordon did his thesis at the Lawrence Livermore Radiation Laboratory and received a Ph.D. in engineering science in September of 1962. He then worked at Aerojet General Nucleonic in San Ramon on small mobile nuclear power plants for the Army and NASA, where he became the vice president and technical director. In June, 1965, he was transferred back to Azusa because of a severe problem with the Snap 8 nuclear program.

In Azusa, he became the Head of Mechanical Operations and was responsible for a space nuclear power system, advanced composite structures, and small free-flooded submarines used in naval covert operations. This is when he learned to scuba dive.

With a new Aerojet management installed in 1970, he did not see any future for mechanical operation in an electronics division. After much discussion, he and two associates purchased the Advanced Composite Division and left Aerojet in May of 1970 to become entrepreneurs. The division was renamed Structural Composites Industries (SCI).

The goal of SCI was to convert aerospace technology to commercial products. Solid rocket motor cases were redesigned to be high pressure gas containers. SCI went from 95% aerospace to 90% commercial and is today, the world's major supplier of light weight, high pressure composite gas cylinders. Operating in the commercial world was exciting, and being under-financed was very challenging. An excellent offer induced the sale of SCI in May, 1983.

Retired and recovering from open heart, by-pass surgery in August, 1984, he turned his attention to writing a book of ethics based on science.

www.ingramcontent.com/pod-product-compliance
Lightning Source LLC
Chambersburg PA
CBHW081130170526
45165CB00008B/2620